Vibrant Harvest

Quarto.com

First Published in 2026 by Cool Springs Press,
an imprint of The Quarto Group,
100 Cummings Center, Suite 265-D, Beverly, MA 01915, USA.
T (978) 282-9590 F (978) 283-2742

EEA Representation, WTS Tax d.o.o.,
Žanova ulica 3, 4000 Kranj, Slovenia.
www.wts-tax.si

30 29 28 27 26 1 2 3 4 5

ISBN: 978-0-7603-9511-0

Digital edition published in 2026
eISBN: 978-0-7603-9512-7

Library of Congress Cataloging-in-Publication Data is available.

Design and page layout: Sarah Pyke
Cover Image: Kami Arant, kamiarant.com
Photography: Sandra Mao, except pages 6, 9, 12, 30, 48, 70, 86, 108, 128, 163 (Kami Arant, kamiarant.com)
Illustration: Shutterstock

Printed in Guangdong, China TT082025

Vibrant Harvest

Cultivating a Kaleidoscope of Colors in Your Vegetable Garden with Heirlooms, Modern Hybrids, and More

Sandra Mao

CONTENTS

welcome
FALL

INTRODUCTION

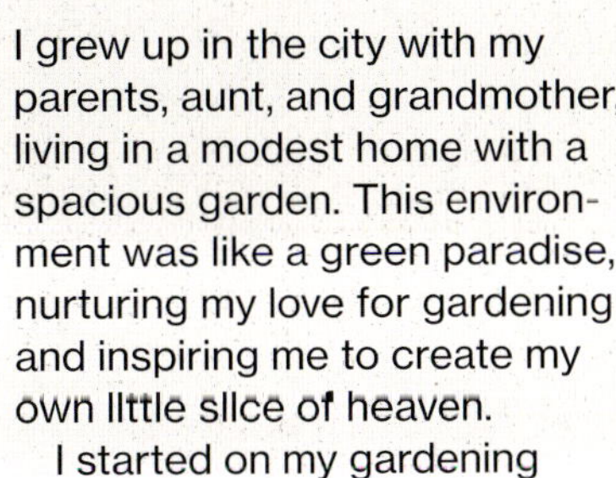

I grew up in the city with my parents, aunt, and grandmother, living in a modest home with a spacious garden. This environment was like a green paradise, nurturing my love for gardening and inspiring me to create my own little slice of heaven.

↑ The end of the summer season harvest

I started on my gardening journey with a few containers in the corner of my backyard. I still vividly recall the moment I harvested my first tomato—which brought me so much joy and happiness! This successful gardening experience made me want to expand my garden to a whole new level. In the autumn of that same year, I started my Instagram account, "Sandra Urban Garden," never anticipating that I would reach today's level of success. Over the years since that first tomato, I have expanded my garden, incorporating more and more raised beds—now I have a total of twenty.

Gardening and growing plants provides opportunities for great satisfaction, enjoyment, and joy. And not only that—a garden should be a feast for the eyes too. The first color that naturally comes to mind when thinking about gardening is, of course, green. But what about all the other wonderful colors? Vegetable gardens absolutely don't have to be boring plots of green!

When it comes to creating a garden, I am always thinking about color. The colors of your garden represent your energy. The color in your garden can have an immediate impact on your mood. I love different colors of flowers incorporated with different colors of vibrant vegetables. The more colors, the better! Growing colorful vegetables is a bright addition to your garden, and different colors of vegetables have different health benefits when they are eaten.

Did you ever hear people say, "Eat the rainbow?" I remembered as a kid, my mom always said, "Eat your vegetables, because they will make you strong and smart. And eat more carrots because they are good for your eyes." Well, Mom is always right!

Rainbow vegetables are a nutrient-dense food that offer a range of health benefits when incorporated into a balanced diet. They are particularly rich in vitamins A, C, and E, as well as iron, calcium, potassium, and fiber. Purple kale, rainbow Swiss chard, and carrots are excellent sources of vitamin A. Rainbow and other varieties of tomato are especially rich in vitamin C.

About This Book

Whether you are a beginner, an amateur, or a professional gardener, this book will serve as a comprehensive resource, providing essential knowledge and inspiration so you can cultivate and design your dream garden. My primary purpose in writing this book is to help you in developing a visually stunning and bountiful garden that offers an array of delectable flavors.

In this book, I have included many tips, suggestions, and pieces of advice to help you successfully navigate your vegetable-growing journey. These include the following items:

- A step-by-step guide to growing forty-two colorful vegetables, herbs, and flowers
- Tips on how to start and build your own vibrant vegetable garden
- Advice on how to choose a color palette for your garden
- Information to help you choose the type of garden best suited to your site, create healthy soil, and apply the right soil amendments
- Guidance on how you can grow from seeds—like a pro
- Recommendations about how to prevent pests and disease in the garden and how to treat them organically
- Instructions on how to preserve all your harvests, including some bonus recipes

↑ Our beloved Lucky in the south side garden in fall. Remembering Lucky (2013–2025).

↑ Springtime in the south-side garden

1

MEET THE VEGGIES

It is almost spring, the beginning of the growing season, and you are dreaming of sunshine and thinking of growing your own food. You are excited and start to map out a garden plan, ordering lots of seeds and buying a lot of plants. Then, you get to feeling like the hype is gone, because you don't know where to start, how to grow from seed, or what crops to grow first. On the horizon, there's also the challenge of how to battle pests and diseases. And while there's tons of information available out there, sifting through it can be overwhelming.

Growing your own food will give you satisfaction and enjoyment, but sometimes it is not an easy process. Trust me, growing vegetables is fun, but know that it will also take time for it to be fun. There is a learning curve. Remember, Mother Nature never rushes.

Gardening is a very rewarding hobby. Follow the guidance in this book and before you know it, you'll be enjoying the best and most unique vibrant vegetables and fruits you've ever eaten. You will be amazed by their sweet taste and juicy flavors. There is absolutely nothing that compares to fresh vegetables harvested from a garden—that you grew yourself.

Starting small is the best way for a beginner to dive in. Make a list of the things that need to be done and set priorities for the items in that list. Starting small also gives you more time to observe and care for each plant so you can have healthier growth and higher yields. Choose a few easy crops to begin with or choose a handful of crops that you and your family enjoy eating the most. And remember, don't worry—I got you!

This part is designed to give you a place to start and to provide focused, in-depth information about individual crops so you can avoid being frustrated and overwhelmed. This part of the book will also provide you with straightforward, step-by-step, seed-to-harvest growing instructions for many types of vibrant vegetables, herbs, and even some edible flowers. For each I have included growing advice, pest and disease control and prevention tips, harvesting and storage information, and insights on the most popular varieties of each vegetable.

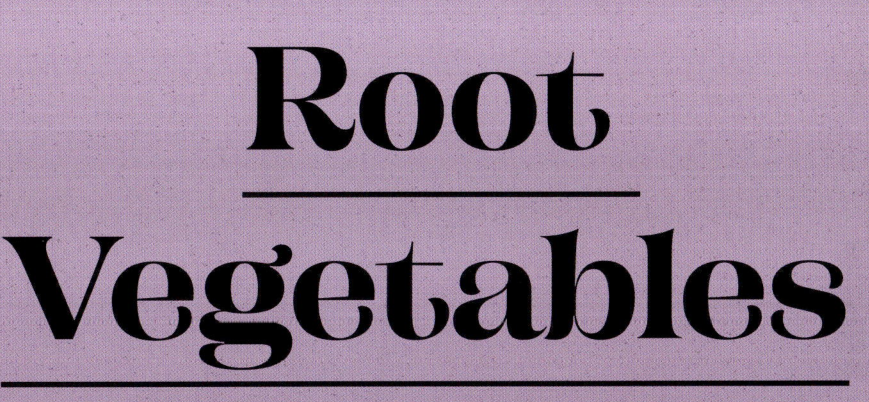

Root Vegetables

RAINBOW BEETS

Rainbow beets are a relatively easy-to-grow vegetable renowned for their earthy, sweet, and tangy flavor. The edible green tops are also highly prized for their delicious taste. Notably, the greens are particularly rich in iron compared to other vegetables. Beets are a nutrient-dense food, packed with antioxidants, vitamin A, vitamin C, fiber, and iron.

While the traditional color of beets is red, this plant exhibits a range of root colors, including yellow, white, and orange. Beets belong to the same plant family as Swiss chard and spinach.

Beets exhibit remarkable growth potential, thriving in full sun or partial shade, and have good resistance to pests and diseases. A unique characteristic of beets is their ability to withstand frosty temperatures approaching freezing, making them an ideal choice for northern gardeners to extend their harvest season.

Planting

Rainbow beets are a cool-season crop that thrives during early spring and fall weather. They can be started indoors or direct-sown outside in early spring to extend their growing season. Beets, because of their shallow roots, also grow well in raised beds and containers. Beet seeds are usually held together in a cluster that can easily result in three to four seedlings growing close together that will need thinning. Beets prefer nutrient-rich, loose soil with good drainage.

Direct-Sowing

Sow beet seeds outdoors as early as February. To expedite germination, presoak the seeds. Sow the seeds ½ inch (1 cm) deep and 3 to 5 inches (7.5–13 cm) apart. Cover the growing area with row cover fabric or fine mulch to reduce weed competition. Ensure that the soil remains moist during the germination period.

Sowing Indoors

Sow beet seeds indoors early, especially in colder regions. Fill a seed tray with seed-starting mix and sow one seed per cell, ensuring the seed is at a depth of 1 inch (2.5 cm). Be aware that beet seeds germinate in clusters, resulting in multiple seedlings per cell. You can thin them out later. Maintain consistent moisture during the germination period. Beet seedlings are ready for transplanting to a larger pot or outdoors when you see a couple of sets of true leaves. The first set of seedling leaves are called "cotyledons," and they won't look like the plant's actual leaves. Later sets of leaves will resemble the plant, and they are called the "true leaves."

Container Planting

Rainbow beets can easily be cultivated in containers. The container should have a depth of at least 12 inches (30 cm) and be filled with well-draining soil. The diameter of the container will determine the number of beets you can plant. Larger beets should be planted about 5 inches (13 cm) apart in a container.

Care

Water

During the germination period, maintain consistent soil moisture, but avoid waterlogging. Once established, beets require minimal watering, preferring moderately moist soil. It is essential to avoid overwatering, as this can promote excessive leafy growth. For in-ground cultivation, water once a week. For raised-bed gardening, water 1 to 2 times a week. For container gardening, water more frequently. Use a mulch to enhance soil moisture retention.

↑ Thin so that beets are at least 5 inches (13 cm) apart if growing the larger sized types.

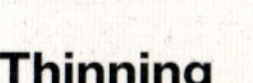

Thinning

Thinning is essential for promoting the development of larger roots. The spacing between roots will directly correlate to their ultimate size. The optimal spacing for beets is approximately 5 inches (13 cm), although I opt for a wider spacing of 8 inches (20 cm) to achieve the most substantial size for my harvest.

Fertilizer

Beets are heavy feeders during the growing period, so knowing how to fertilize them is very important. Use a fertilizer that is high in potassium and phosphorus but lower in nitrogen.

Weeds

Beets do not compete well with weeds. Weeds can cause severe root disturbance to the beet roots during the germination and establishment period. Therefore, it is advisable to gently remove weeds early. Regular weed control is necessary to reduce competition with the plant for nutrients and water. Hand weeding is recommended to avoid damaging the roots.

Pests and Diseases

The most common pests that attack beets are leaf miners and flea beetles. The most common disease is leaf spot. (See page 152.)

Companion Plants

Beets grow well with garlic, carrots, kale, Brussels sprouts, mint, radishes, bush beans, lettuce, cabbage, broccoli, sage, and marigolds. Avoid growing beets with these plants: spinach, chard, and pole beans.

↑ These beets have reached their maturity date and are at a size of about 3 to 4 inches (7.5–10 cm) in diameter.

Harvest and Storage

Beets typically mature within a 50- to 70-day period, depending on the specific types. Unlike other vegetables, beets offer flexibility in harvesting, allowing for the selection of a range from young to fully mature roots. Consequently, if beets are cultivated in the fall, they should be ready for harvest before the ground freezes.

When to Harvest

Beets are ready for harvesting when the top shoulder of the roots grows 1 to 2 inches (2.5–5 cm) above the soil. Alternatively, you can determine readiness by their size, which should be comparable to a tennis ball.

Growing beets offers a dual benefit; while I patiently wait for the roots to mature, I can savor their delectable leaves. Beet leaves are edible, but it is crucial to avoid excessive harvesting of the foliage, as beets require a healthy green top to grow a larger root. I intentionally leave a few roots in-ground for several months to continue to harvest the leaves.

How to Store

Before you store beet roots, remove the leaves, but do not discard them. You can incorporate beet leaves into a salad bowl, stir-fry, or even add them to your morning juice.

Beets are exceptionally flavorful and can be incorporated into various dishes. I particularly enjoy roasting the roots with olive oil and sprinkling them with sea salt. If you have surplus beets, you can cut them into smaller pieces, boil them, and store them in the freezer or preserve them by canning.

My Favorite Vibrant Varieties

- **'CHIOGGIA' BEETS:** These can be big. The roots are red on the outside, but the inside has red and white concentric rings.
- **'CROSBY EGYPTIAN' BEETS:** This beet has a flattened shape and a sweet but slightly earthy flavor.
- **'DETROIT DARK RED' BEETS:** One of the most popular types of beets, it is round and sturdy, with red roots.
- **GOLDEN BEETS:** These beautiful golden roots have sweet, yellow flesh.

GARLIC

Garlic, a member of the Allium family, which also includes chives, is a versatile ingredient commonly used in culinary creations. Planted in the autumn season, garlic is relatively straightforward to cultivate and demands minimal attention. Once planted, the cloves transform into bulbs within a timeframe of approximately 7 to 9 months. Given its widespread use in the kitchen, garlic seamlessly integrates into a variety of dishes, enhancing their flavor and aroma.

Garlic is classified into two main types:

Softneck Garlic: This variety is characterized by its production of larger bulbs with a milder flavor profile. Notably, it lacks the hard center stem in the bulb commonly referred to as a "scape." The plant stem retains its softness after harvest, enabling it to be braided for storage or decorative purposes. Softneck garlic flourishes in warmer climates and is the favored choice for gardeners in southern regions. Furthermore, it exhibits an exceptionally extended storage life. The majority of supermarket garlics are the softneck type.

Hardneck Garlic: This garlic variety is named after the hard, central bulb stalk known as the "scape." When planting, removing the scape is essential to encourage the bulbs to grow and develop to their full size. Hardneck garlic has a shorter storage life compared to the softneck varieties, but it exhibits exceptional cold hardiness, making it the preferred choice for northern gardeners.

↑ Garlic cloves ready for plantings

Planting

Garlic is primarily planted in the fall for northern regions. Plant garlic cloves approximately 6 weeks before the first frost date. By planting garlic cloves early in the fall, you will allow sufficient time for them to develop healthy roots before the ground freezes.

Garlic cloves will not grow during the winter, but they will begin to sprout when spring arrives.

Garlic can also be planted in early spring in warmer, southern regions. However, garlic cloves require at least a couple of months of chilling time. Therefore, it is recommended to order garlic cloves early and store them in the refrigerator until planting day. After planting, you will observe new garlic sprouts within 1 week.

Garlic cultivation can be done in-ground, or bulbs can be grown in raised beds or containers. Garlic thrives in full sun with a well-drained, loose, and fertile soil. Unlike other plants, garlic can only be propagated from cloves. A single clove will grow and ultimately develop into a complete bulb.

1. Prepare the soil by incorporating a few inches of organic compost and applying an all-purpose fertilizer.

2. Separate the garlic cloves but refrain from peeling off the outer skin paper. It is advisable to purchase garlic cloves specifically intended for planting rather than using those from grocery stores, as the latter may have been treated to prevent them from sprouting.

3. Plant the garlic cloves 1 to 2 inches (2.5–5 cm) deep and 5 to 6 inches (13–15 cm) apart. Ensure that the pointed end is facing up, as this is the site of shoot development. Cover the cloves lightly with a thin layer of soil.

4. Apply a layer of mulch, such as straw, grass clippings, or dried leaves, that is 3 to 4 inches (7.5–10 cm) deep to help conserve moisture and regulate temperatures.

Care

Water

During the winter months, garlic does not need excessive watering. Water is only required after planting, and watering should be suspended when the ground freezes. Watering can resume once the temperature rises above freezing again. It is imperative to avoid overwatering, as this can lead to bulb rot. Conversely, underwatering can stress the plant. So, during the dry months, apply 1 inch (2.5 cm) of water to the plant once per week. When the leaves begin to turn yellow and harden, it is time to cease watering to allow the bulbs to mature.

Fertilizer

Garlic is a heavy feeding crop. In the spring or after the garlic cloves sprout, provide them with bone meal and a nitrogen source every 3 to 4 weeks.

Pests and Diseases

The most common pests and diseases for garlic are thrips, white rot, leaf miner, and onion maggots. (See page 150.)

Companion Plants

Garlic grows well with tomatoes, peppers, potatoes, spinach, carrots, beets, cabbages, eggplants, broccoli, kale, dill, and marigolds. Avoid growing garlic with melons, asparagus, peas, and beans.

Harvest and Storage

Garlic harvesting season varies depending on the region and the time of planting. In most areas, garlic is typically ready for harvest from late June to August.

↑ Delicious garlic scapes

How to Harvest Garlic Scapes

If you are growing hardneck garlic, a flowering stalk emerges from the center of the bulbs. This typically occurs in early summer. These scapes must be harvested because you want the plant's energy to concentrate on bulb development, resulting in larger harvests. Garlic scapes are edible and can be considered a bonus crop. Some scapes will grow straight, while others will curve. Allow the scapes to grow longer and then cut or snap them as close to the base as possible.

How to Harvest Garlic Bulbs

When you observe the bottom garlic plant leaves turning yellow and the tops falling over, this is an indication that your garlic is ready for harvest. Typically, this occurs approximately 1 month after harvesting the garlic scapes. If you are uncertain, you can pull on a plant to ascertain if the bulb has matured. However, it is advisable to refrain from hand pulling the garlic, as this can damage the bulbs, encourage bulb rot, and result in a shorter storage time. Instead, use a pitchfork to lift the soil surrounding the plant and then gently extract the bulbs.

How to Store

The final step in the garlic harvesting process is curing the bulb for storage. It is crucial to refrain from washing the garlic bulbs, as this can lead to mold growth. Instead, place the bulbs in a dry and well-ventilated area. Allow them to dry for approximately 1 month. Once cured, you can store the garlic in wooden crates, wire baskets, or wire shelves. Softneck garlic can also be braided for storage.

My Favorite Vibrant Varieties

- **HARDNECK GARLIC VARIETIES:** 'Chesnok Red', 'Music', 'German White', 'Russian Red', 'Purple Glazer', 'Rocambole', 'Purple Stripe'.
- **SOFTNECK GARLIC VARIETIES:** 'Elephant', 'Early Italian', 'California Early', 'Inchelium Red', 'Transylvania', 'Italian Late', 'Texas Rose', 'Viola Francese'.

RAINBOW CARROTS

Rainbow carrots are a cool-season crop and one of the most straightforward root vegetables to cultivate. They thrive in sunny areas and exhibit a range of sizes, types, shapes, and colors—including purple, yellow, white, and multicolored. While homegrown carrots may not always be as straight as supermarket carrots, their taste and sweetness are significantly superior. Carrots are low in calories and are an exceptional source of vitamin A.

Planting

In warmer climates, early rainbow carrot crops can be sown in February. In colder regions, mid-spring is the preferred time for sowing, after the last frost date has passed. The optimal soil temperature range for carrot growth is 55°F to 70°F (13°C–21°C). Carrots thrive in full sun, with fertile, deep, loose, and well-drained soil.

For optimal results, direct-sow carrot seeds into in-ground beds. Carrots can also be grown in raised beds or containers. Plant carrot seeds at a depth of ¼ to ½ inch (0.6–1 cm) and approximately 4 to 6 inches (10–15 cm) apart in each row. Spacing is not critical at this stage, as thinning can be performed later. Carrots don't need excessive manure or fertilizers, as these can encourage the development of multiple side roots instead of nice long, tapered roots. Carrots dislike being disturbed or transplanted, so sow them directly where you want them to grow.

Care

Water

Carrots need adequate soil moisture for optimal quality and size. Insufficient water will result in a bitter taste and a tough root. If they are grown in heavier soils, water them deeply once or twice per week. If they are growing in raised beds or containers, water them 2 or 3 times a week. Alternatively, you can perform the finger test to determine your soil moisture levels. Poke your finger into the soil, and if the top layer of soil is dry 1 to 2 inches (2.5–5 cm) down, it's time to water. Applying a mulch will also help in retaining moisture and preventing direct sunlight from striking the roots.

Thinning

One of my preferred techniques for growing carrots is to thin out the seedlings. For most carrot varieties, the recommended spacing is 2 to 4 inches (5–10 cm) apart. Thinning carrots means removing some of the seedlings to allow the remaining ones to have more space and nutrients for growth and development. During the initial thinning process, when the seedlings have grown approximately 2 to 3 inches (5–7.5 cm) tall, identify the weaker seedlings and gently remove them. To minimize disturbance to the surrounding roots, you can also cut the seedlings off at the base of the plant. Repeat this process every 4 to 5 weeks, as more seedlings will emerge during this period.

Fertilizer

Carrots are not a heavy feeding crop, so it is not necessary to fertilize them often. However, it's okay to fertilize them a couple of times during the season, but use

↑ The ideal spacing for carrots is 2 to 4 inches (5–10 cm) apart.

a fertilizer that is low in nitrogen, otherwise the nitrogen will promote foliage growth at the expense of roots.

Weeds

Carrots do not compete well with weeds. Weeds can cause severe root disturbance to carrots during the germination and establishment period—so pull them early and gently.

Pests and Diseases

The most common pests and diseases that attack carrots are carrot rust flies, root-knot nematodes, and leaf blight. (See page 150.)

Companion Plants

Carrots grow well with marigolds, oregano, cilantro, tomatoes, lettuce, onions, bush beans, radishes, and beets. Avoid planting carrots with dill, parsnip, potatoes, fennel, and celery.

Harvest and Storage

Generally, carrots are mature in 70 to 100 days. But remember, the smaller carrots usually taste better. Carrots are biennial, so if you leave the plant in the garden and the top on to flower, it will produce seeds the next year.

When to Harvest

Carrots are ready for harvesting when the top of the root is slightly above the soil level and is approximately 1 to 2 inches (2.5–5 cm) in diameter. If you are uncertain, you can simply pull a few roots to determine their readiness. For smaller carrots, gently wiggle the root and pull them out. However, for larger carrots, you may need to use a garden fork to loosen the soil around them and gently pull them upward.

How to Store

After you have harvested your carrots, cut off the top leaves and save them for eating. You can make a delectable pesto out of the carrot leaves. I enjoy adding carrot top pesto to chicken noodle soup. Carrots are exceptionally flavorful and can be incorporated into various dishes, whether enjoyed raw or cooked. If you have surplus carrots, cut them into smaller pieces and store them in the freezer or preserve them by canning.

My Favorite Vibrant Varieties

- **'DANVERS' CARROTS:** These heirlooms grow up to 6 to 8 inches (15–20 cm) long.
- **'LITTLE FINGER' CARROTS:** These fast-growing baby carrots are sweet and tender. They are best for snacking.
- **'MANPUKUJI' CARROTS:** These carrots can grow up to 3 feet (91 cm) long. Yup, you read it right—so much fun to grow and they taste amazing.
- **'NANTES' CARROTS:** These slender 6 inch (15 cm)-long carrots with a rounded end, great crunch, and super sweet.
- **'PURPLE DRAGON' CARROTS:** These carrots with a beautiful red-purple exterior and bright yellow-orange inside.

While homegrown carrots may not always be as straight as supermarket carrots, their taste and sweetness are significantly superior.

RAINBOW RADISHES

Rainbow radishes are annual vegetables that are exceptionally easy to cultivate. Radishes as a whole exhibit a diverse range of sizes, varieties, and colors. They are characterized by their rapid growth rate, with most varieties reaching maturity within approximately 1 month after planting. This makes them an ideal vibrant vegetable choice for beginners. Additionally, radishes possess some cold tolerance and belong to the large Brassicaceae (or cabbage) family. Other brassicas include bok choy, broccoli, Brussels sprouts, cabbage, cauliflower, collards, kale, kohlrabi, and turnips.

In terms of nutritional value, radishes are low in carbohydrates but high in antioxidants, vitamin C, and other minerals such as calcium and potassium. Notably, the green leafy portion of the radish plant is also edible and possesses a delightful flavor.

Planting

Rainbow radishes require full sun to partial shade, meaning at least 4 to 6 hours of sunlight per day. However, excessive shade can hinder the development of roots in favor of mostly green foliage.

Rainbow radishes (and most other types) are straightforward to cultivate from seeds and typically germinate within 3 to 6 days. Sow seeds during early spring or late summer. Rainbow radishes can be grown directly in the ground, and in raised beds or containers that have been filled with loose, fertile, and well-draining soil. To ensure a continuous harvest, sow seeds every 2 weeks throughout the growing season.

Direct-Sowing

Rainbow radish seeds can be direct-sown outdoors from spring through fall. Sow seeds 3 to 5 weeks before the last spring frost date. Sow seeds ½ to 1 inch (1–2.5 cm) deep and 1 to 3 inches (2.5–7.5 cm) apart.

↑ Easter Basket radishes

↑ Colorful radishes growing in a raised bed

For fall planting, sow seeds 3 to 5 weeks before the first fall frost. Cover the area with row cover fabric or mulch to protect the seeds from frost.

Sowing Indoors
Sow rainbow radish seeds indoors early, particularly in colder regions. Fill the seed tray with seed-starting mix and sow 1 to 2 seeds per cell, ensuring they are planted 1 inch (2.5 cm) deep. Maintain even soil moisture during the germination period. Seedlings are ready for transplanting when you observe approximately two sets of true leaves.

Radishes as a whole exhibit a diverse range of sizes, varieties, and colors.

Care

Water
Rainbow radishes (like other types) require consistent watering. Maintain a consistent soil moisture around the plants, ensuring it is neither excessively wet nor dry. Excessive water can lead to radish roots splitting or cracking, while allowing the soil to dry out can cause the plant to bolt rapidly and flower. Mulching can help in soil water retention.

Thinning
Thinning is crucial for achieving full-sized radishes to harvest. Overcrowding hinders growth, so when the plants reach 2 to 3 inches (5–7.5 cm) tall, selectively and gently remove some or sever the stems at soil level. Leave a space that is two fingers wide (approximately 2 inches [5 cm]) between them so the remaining radishes can grow unimpeded. Do not discard the seedlings; they can be cleaned and incorporated into your next salad.

Fertilizer
Rainbow radishes are not a heavy feeding crop and they grow quite rapidly, so just apply an all-purpose fertilizer at planting, and consider applying another application with fish fertilizer a few weeks before harvest. Avoid applying excessive nitrogen-based fertilizers, as this can promote leafy growth rather than root growth.

Weeds
Rainbow radishes are not well-suited for competition from weeds. Weeds can cause significant root disturbance during the germination and establishment periods. To mitigate this, it is advisable to gently remove weeds early. Regular weed con-

trol is necessary to reduce competition for nutrients and water. Additionally, hand weeding is recommended to avoid damaging the radish roots.

Pests and Diseases

Common pests of radishes are slugs, snails, aphids, flea beetles, and cabbage loopers. (See page 150.)

Companion Plants

Rainbow radishes grow well with lettuce, carrots, tomatoes, peppers, mint, dill, bush beans, peas, eggplants, cucumbers, and oregano. Avoid growing radishes with brassicas, melons, pumpkins, sunflowers, and potatoes.

Harvest and Storage

Harvest rainbow radishes (and other types) early for the best taste. If you leave them in the ground too long, they will become woody.

When to Harvest

There are several methods to determine when rainbow radishes are ready for harvest. Firstly, refer to the seed package for the specific maturity dates, which show the approximate time from seed planting to harvest. Alternatively, select a few radishes and taste them. Another indicator is when the roots reach a diameter of approximately 1 to 2 inches (2.5–5 cm), depending on the type.

How to Store

After harvesting, separate the leafy portion of the radish, but do not discard it. You can incorporate the leaves into soup or prepare a delectable fresh kimchi with them. Rinse and pat the roots dry. Wrap them in a paper towel and place them in a plastic bag. Store them in the refrigerator for a few weeks. Additionally, you can preserve them by making pickled radish or kimchi. (See page 169.)

My Favorite Vibrant Varieties

- **EASTER BASKET MIX RADISHES:** A beautiful mix of colors and different radish shapes
- **'FRENCH BREAKFAST' RADISHES:** A good radish that is heat tolerant, has an oblong shape with white flesh at the end, and a mild flavor
- **'WATERMELON' RADISHES:** Very beautiful, with white-green skin, bright pink flesh, and a mild flavor
- **'WHITE ICICLE' RADISHES:** This is a fall radish of medium size, with a mild flavor and good crunch.

ONIONS

Onions are a simple and rewarding crop to cultivate due to their culinary value. They offer a range of types, each with its unique characteristics, including pest and disease resistance.

Onions exhibit a diverse spectrum of colors, including those bulbs that are white, yellow, and red. Each color possesses distinctive flavor profiles, ranging from mild to spicy and sweet.

The structure of an onion is closely related to its appearance. Each leaf of the onion corresponds to a single onion ring. Larger leaves indicate larger onion rings, while multiple layers are formed by the accumulation of leaves.

Planting

Onions are a cool-season crop that can be propagated through seeds, sets aka small onion bulbs, or transplants. Onions thrive in sunny locations with loose, well-drained, and nutrient-rich soil.

Selecting the appropriate onion varieties for your location is paramount. Onion varieties that form bulbs are influenced by day length and the number of daylight hours.

There are three primary categories of onions:

- **Short-day onions** are suitable for southern latitudes, requiring approximately 10 to 12 hours of daylight.
- **Long-day onions** are suitable for northern latitudes, requiring approximately 14 to 16 hours of daylight.
- **Intermediate-day onions** are suitable for central latitudes, requiring approximately 12 to 14 hours of daylight.

Transplants

Plant the onion seedling transplants 4 to 6 weeks before the last frost. Plant the onion seedlings 1 inch (2.5 cm) deep and about 4 inches (10 cm) apart.

Seeds

Onions can be propagated from seeds, although this method requires a longer timeframe and is not recommended for beginners. For indoor cultivation, onion seeds can be sown as early as January. A heat mat may be necessary beneath the seedling tray, as onion seeds require temperatures between 60°F and 70°F (16°C–21°C) for germination. Additional grow lights may be required after germination indoors.

↑ Onions can be grown in raised beds, which offer better drainage, are easier to maintain, and have fewer weeds.

Sets

Growing onions from sets is the easiest and most cost-effective way to do it. Choose smaller sets because they're less likely to bolt; ultimately, they will give you bigger bulbs. Bigger sets bolt way too fast. Bolting occurs when plants are triggered to flower and set seed too early, making the vegetables themselves less desirable.

Plant onion bulbs early in the spring when the ground isn't frozen. In warmer regions, plant them 4 to 6 weeks before the first fall frost. Plant the bulbs with the pointy end up, about 1 inch (2.5 cm) deep, and space them about 4 inches (10 cm) apart. Cover the planting area with mulch to help keep the soil moist.

Care

Water

Onions require consistent watering to help bulb expansion. While deep watering is unnecessary due to their shallow root system, approximately 1 inch (2.5 cm) of water per week is sufficient. Overwatering can lead to bulb rot, so it is crucial to avoid excessive moisture. Light mulching can help in retaining soil moisture in drier climates.

Thinning

Thinning is a crucial step in achieving full-sized bulbs. Thin onion seedlings so there are 4 to 5 inches (10–13 cm) between each one for larger bulb production.

Fertilizer

Onions are heavy feeders. Fertilize the plants every 2 to 3 weeks with a high-nitrogen fertilizer to support good leaf growth. Remember, the larger the leaf, the larger the onion ring in the bulb.

Pests and Diseases

The most common pests and diseases for onions are thrips, white rot, and purple blotch. (See page 152.)

Companion Plants

Onions grow well with beets, spinach, brassicas, eggplants, tomatoes, peppers, lettuce, carrots, marigolds, and dill. Avoid growing onions with garlic, shallots, and leeks.

Harvest and Storage

When to Harvest

Onion bulbs are typically ready for harvest after approximately 125 to 150 days from seeding or 60 to 80 days from planting sets, depending on the specific variety and growing conditions. If onions are planted in the spring, they should be ready for harvest in late summer, typically when the foliage begins to turn yellow or brown, the necks become soft and droopy, and the tops flop over. To harvest the onions, carefully loosen the soil with a garden fork and gently pull the onion bulbs up to avoid bruising them.

How to Cure

Curing onions is a process that prepares them for long-term storage. During curing, the outer layer of the onion bulbs is dried, which helps to maintain their firmness and freshness for an extended period.

Onion bulbs require ample space for good air circulation and drying conditions, warmth (approximately 75°F [24°C]), and protection from direct sunlight. It is best to lay onions in a single layer for curing. Ensure that you turn the bulbs every few days to help even drying. Once the necks and stems are completely dry, trim the roots and leaves from the bulb.

How to Store

The optimal storage location for onions is in a dry, dark, and cool environment. To maintain their flavor and prevent bulb rot, store onions in wooden crates, wire baskets, or wire shelves.

↑ Harvested red onions ready to go through the curing process

My Favorite Vibrant Varieties

Long-Day Varieties

- **'RED CREOLE' ONIONS:** Medium-size bulbs with a spicy flavor and red flesh
- **'WALLA WALLA' ONIONS:** Onions with huge bulbs, yellow skin, and a mild and sweet flavor
- **'YELLOW SWEET SPANISH' ONIONS:** Bulbs with a big round shape, white and yellow flesh, and sweet with a mild flavor

Short-Day Varieties

- **'RED BURGUNDY' ONIONS:** Sweet with beautiful bright red flesh and flat-shaped bulbs
- **'TEXAS EARLY GRANO' ONIONS:** A sweet heirloom onion
- **'YELLOW GRANEX' ONIONS:** Extremely sweet, perfect for southern gardeners

Intermediate-Day Varieties

- **'CANDY' ONIONS:** An onion with big bulbs and the sweetest flavor
- **'CIMARRON' ONIONS:** One of the most productive yellow onions, with large globe-shaped bulbs and a sweet and mild flavor
- **'STOCKTON RED' ONIONS:** An onion with pink and white flesh, crisp and sweet
- **'SUPER STAR' OR 'SIERRA BLANCA' ONIONS:** A beautiful white onion with globe-shaped bulbs

MULTICOLORED POTATOES

Multicolored potatoes are a simple crop to cultivate under minimal care, providing gardeners with an enjoyable and rewarding harvest. They come in various varieties, sizes, and colors—including yellow, orange, red, white, and purple. All potatoes are an energy-dense crop because they are a high-quality source of carbohydrates. They are free of fat, gluten, and cholesterol. Additionally, they are an excellent source of vitamin C, vitamin B6, and potassium.

Planting

Multicolored potatoes (and other types) are a warm-season crop that can be planted in early spring. Alternatively, they can also be grown as a winter crop in warmer climates. Multicolored potatoes thrive in full sun and loose, fertile soil. They can be grown in-ground, as well as in raised beds, containers, or grow bags. Multicolored potatoes are grown from seed potatoes—small potatoes or pieces of potato with at least one eye. Remember that while the potatoes themselves are a delectable food, their leaves are toxic.

The first crucial step in growing multicolored potatoes (as well as other types) is to locate disease-free seed potatoes. It is advisable to purchase certified disease-free seed potatoes. I would not recommend growing potatoes from the grocery store, as these cooking potatoes are often treated with growth inhibitors to prevent sprouting.

There are two types of potatoes:

Determinate Potatoes:

- Are fast maturing, with harvests in approximately 90 days
- Do not require mounding and need more space to grow
- Cease production of new tubers when the plant dies or flowers appear

Indeterminate Potatoes:

- Mature in 120 to 160 days
- Require minimal space for growth and can be grown in containers
- Must be mounded as the plant grows
- Continuously produce new tubers until frost arrives

Mounding

Hilling or mounding refers to the practice of creating a mound around the base of a potato plant. The primary objective of hilling potatoes is to maintain darkness around the tubers. Exposure to sunlight causes the green pigment on the skin to turn toxic.

When the multicolored potato plants have grown to approximately 8 to 10 inches (20–25 cm) tall, additional soil is added to the mound. This process is repeated several times as the plant continues to grow.

Direct-Sowing

If you are growing in-ground, add organic compost and all-purpose fertilizer. Cut the seed potato into pieces with at least two eyes, as these are the points where the shoots will grow. After cutting, cure the potato tuber by leaving it laying out exposed to the air for 1 to 2 days to allow the skin to dry. This process helps prevent the tuber from rotting when the soil becomes wet.

Place the potato in the bottom of a trench with the eyes facing up, and cover it with soil.

Container Planting

If you are using a container or grow bag, ensure it has adequate drainage, and a capacity of 12 to 15 gallons

(45–57 L) and a height of no more than 3 feet (91 cm).

For raised-bed gardening, prepare a well-draining potting mix by combining high-quality potting soil, organic compost, and organic all-purpose fertilizer.

Care

Water

Multicolored potatoes require consistent moisture throughout their growth period, but they dislike soggy soil. Mulching can help in maintaining moisture retention.

Multicolored potatoes (and other types) require approximately 1 to 2 inches (2.5–5 cm) of water per week. If you can, water around the base of the plants and not on the leaves. During the summer months, especially when flowering begins and after the flowering stage, the potatoes require additional water. During the flowering stage, the plants produce tubers, so consistent watering is crucial. However, once the foliage turns yellow (about 1 month before harvest), reduce the water supply. This will help curing and prepare the potatoes for harvest.

Fertilizer

Multicolored potatoes (as well as other types) require a comprehensive nutrient regimen throughout their growth cycle to yield substantial crops. Specifically, they need higher levels of phosphorus and potassium.

However, for the development of larger tubers, a subsequent application of nitrogen fertilizer, such as a 5-10-10 (NPK) formulation, is recommended after the plant has reached maturity.

↑ Potatoes growing in a raised bed

Pests and Diseases

The most common pests and diseases of multicolored potatoes are slugs, aphids, cutworms, and late/early blight. (See page 150.)

Companion Plants

Multicolored potatoes grow best with horseradish, garlic, onions, peas, beans, corn, lettuce, spinach, radishes, cilantro, parsley, cabbages, and cucumbers. Avoid planting potatoes with tomatoes, peppers, eggplants, tomatillos, and okra.

Harvest and Storage

Harvesting New Potatoes

All multicolored potatoes can be harvested as new potatoes. These can be harvested in late spring or early summer. New potatoes are harvested when the plants are young before they reach maturity, while the tuber is still small and has thin skin. The thin skin makes the potato extremely tender, succulent, and delicious—but it has a short storage life.

Harvesting Storage Potatoes

Multicolored potatoes for storing are harvested once the plant has fully matured at the conclusion of the growing season. The tubers are ready for harvesting when the foliage turns yellow and becomes dry. This waiting period is necessary to allow the potato skins to thicken, which is optimal for long-term storage. Exercise caution when digging the potatoes to avoid piercing the tubers. Use a garden fork to loosen the soil around each plant and then carefully sift through the soil to find the tubers.

How to Store

Following harvest, select all the desirable multicolored potatoes that are free from damage caused from piercing the skin or bruising. The optimal storage environment is a cool, dry, and dark location.

My Favorite Vibrant Varieties

- **'PURPLE MAJESTY' POTATO:** This stunning potato with dark purple skin and flesh is good for baking, boiling, or frying.
- **RAINBOW MIX POTATO:** This easy-to-grow mixture of colors features many favorite potato varieties. One pound (0.5 kg) of Rainbow Mix seed potatoes can produce between 10 and 15 pounds (5–7 kg) of harvested potatoes under good growing conditions.
- **'RASPBERRY' POTATO:** This potato has vibrant red skin and stunning red flesh. It has a creamy texture and subtly sweet, earthy taste.

Stem Vegetables

PURPLE ASPARAGUS

Asparagus is a perennial vegetable, meaning the plants will produce spears year after year—and they can live for 15 to 20 years. Therefore, it is crucial to select a suitable location for planting. The edible shoot or stem of the asparagus plant is called the "spear." Asparagus varieties come in green, purple, pink, and white, and the spears are characterized by their crunchy texture, sweetness, and tenderness. As the spears mature, they will eventually grow into a plant with bushy green foliage.

Planting

Asparagus is a cool-season crop that thrives when planted in early spring, as soon as the soil is suitable for cultivation. It prefers a sunny location with at least 8 hours of sunlight daily. Additionally, it requires nutrient-rich and well-drained soil with a pH between 6.5 and 7.0. Purple asparagus can be propagated from seeds or crowns.

Sowing Indoors

Sow asparagus seeds approximately 12 to 14 weeks before your average last frost date in the spring. To speed up the germination, soak the seeds in water overnight. Plant the seeds approximately ½ inch (1 cm) deep and with 1 to 2 seeds per cell. Maintain even soil moisture during the germination period and use a heat mat to ensure temperatures remain within the range of 60°F to 80°F (16°C–27°C). Germination is typically achieved within 10 to 12 days. Subsequently, remove the heat mat and relocate the seedlings under grow lights to prevent them from getting leggy.

Crowns

Cultivating asparagus from crowns is the most straightforward method to start new plants.

Prepare holes or trenches approximately 8 to 10 inches (20–25 cm) deep and 10 inches (25 cm) wide. Spread the roots at the bottom of the trench and cover the crown with a layer of soil about 3 inches (7.5 cm) deep. As the young shoots develop, gradually fill in the holes or trenches until they are level with the soil surface.

Care

Watering

Purple asparagus (and other types) require consistent watering to promote robust root and fern (leaf) development. Young asparagus plants need approximately 1 to 2 inches (2.5–5 cm) of water per week, while mature plants require about 1 inch (2.5 cm) of water per week. To enhance water retention, a layer of mulch should be applied to the soil.

↑ Healthy young asparagus shoots (spears)

Weeds

Purple asparagus is susceptible to competition from weeds, which can hinder growth. In home gardens, the most effective method to prevent damage to asparagus shoots and roots when weeding is to regularly remove weeds by hand. Straw mulch or dried leaves can be used in asparagus patches to smother weeds.

Fertilizer

Regardless of whether you grow purple asparagus in in-ground beds or raised beds, applying fertilizer high in phosphorus and potassium, along with organic compost in the trench when planting, will promote root growth and provide the plant with an adequate nutrient supply as it grows. To maintain soil fertility, annually add organic compost in early spring.

Pests and Diseases

Purple asparagus doesn't have many problems in the garden. The most common pests are cutworms, asparagus beetles, and aphids. Common diseases include asparagus rust and purple spot. (See page 150.)

Companion Plants

Purple asparagus grows well with many other plants such as tomatoes, basil, parsley, marigolds, nasturtiums, coriander, dill, cilantro, eggplants, spinach, and peas. Avoid growing purple asparagus with potatoes and alliums like garlic, onions, and leeks.

Harvest and Storage

When to Harvest

Asparagus spears emerge in the spring, but it is advisable to refrain from harvesting them until the third year after planting. During the first 2 years, the spears should be allowed to grow to fern height and develop a robust root system. In the third year, you should be able to harvest the spears for approximately 2 to 3 weeks.

Harvest the spears when they reach a height of about 6 to 8 inches (15–20 cm) and a thickness of ½ to ¾ inch (1–2 cm). To avoid damaging the spears beneath the soil surface, cut them at the soil level using a knife or scissors.

How to Store

Purple asparagus is most flavorful when consumed within hours of picking. However, it can be stored in a sealed, clear, plastic bag in the refrigerator for up to 1 week. Alternatively, purple asparagus (and other types) can be preserved through freezing or canning for extended storage.

My Favorite Vibrant Varieties

- **'JERSEY GIANT' ASPARAGUS:** This type has an excellent yield with large, dark green spears, with great disease resistance.
- **'PURPLE PASSION' ASPARAGUS:** This asparagus has a good yield and large, dark purple spears.

PURPLE KOHLRABI

Kohlrabi, also known as German turnip, is a distinctive vegetable belonging to the Brassicaceae family. It develops long, leafy foliage from a swollen stem, commonly referred to as the bulb. Kohlrabi has both purple and green varieties, with the stem and leafy greens being edible. This vegetable is a nutritional powerhouse, providing a rich source of vitamin C, vitamin A, calcium, iron, and potassium.

Planting

Purple kohlrabi, a cool-season vegetable, can be cultivated in both autumn and spring. It thrives in full sun, with well-draining, fertile, and loose soil. Kohlrabi is straightforward to grow from seeds and can be direct-sown in the garden or started indoors.

Direct-Sowing

Purple kohlrabi seeds can be direct-sown once the soil is suitable for planting. Sow the seeds at a depth of ½ inch (1 cm) and a spacing of 8 to 10 inches (20–25 cm), or up to 12 inches (30 cm) apart for larger bulb types. If you are planting in the fall, ensure that you protect the seedlings, as they are not frost-tolerant when young. However, the mature kohlrabi can withstand a light frost.

Sowing Indoors

Sow purple kohlrabi seeds indoors early, particularly in colder regions. Begin sowing seeds indoors approximately 6 weeks before the last spring frost date. Fill the seedling tray with a seed-starting mix and sow 1 to 3 seeds per cell, ensuring they are planted ¼ to ½ inch (0.6–1 cm) deep. Maintain even soil moisture during the germina-

tion period and place the seedling tray in a location at room temperature—between 60°F and 70°F (16°C–21°C). If necessary, use a heat mat to provide warmth during germination. Remove the heat mat once the seeds have germinated and place the tray under a grow light to prevent leggy seedlings. Kohlrabi seedlings are ready for transplanting when they have developed a couple of sets of true leaves.

Care

Water

Kohlrabi requires consistent watering due to its shallow root system. Water the plant with approximately 1 inch (2.5 cm) of water per week, ensuring that the soil remains consistently moist. Avoid letting the soil dry out, as this can lead to the development of woody and tough bulbs. Instead, water the plant at the base rather than over the leaves, which helps prevent diseases such as powdery mildew and leaf spots. Additionally, applying a layer of mulch can help in retaining moisture and controlling weeds.

Fertilizer

Kohlrabi thrives in nutrient-dense soil. By incorporating a substantial amount of compost during planting, the need for excessive fertilizer can be mitigated. Fertilize kohlrabi with a nitrogen-based fertilizer or an all-purpose fertilizer every 4 to 6 weeks.

Pests and Diseases

The most common pests and diseases for purple kohlrabi are cabbage loopers, aphids, cutworms, powdery mildew, and leaf spots. (See page 150.)

Companion Plants

Purple kohlrabi grows well with beets, bush beans, celery, cucumbers, onions, nasturtiums, lettuce, and herbs. Avoid growing kohlrabi with strawberries, pole beans, peppers, and tomatoes.

Harvest and Storage

When to Harvest

Purple and white kohlrabi exhibit a diverse range of varieties and sizes, with maturity periods spanning 45 to 75 days. They are ready for harvest when the stem (bulb) attains a diameter of 3 to 5 inches (7.5–13 cm). Generally, kohlrabi is best harvested when young and when it is a smaller size.

How to Store

Purple kohlrabi can be stored in the refrigerator for several weeks, or you can make vegetable pickles or kimchi for a longer storage life. (See page 169 for recipes.)

My Favorite Vibrant Varieties

- **'AZUR STAR' KOHLRABI:** This open-pollinated kohlrabi with purple skin and white flesh. Matures in 48 days from seed.
- **'BLAUER SPECK' KOHLRABI:** Frost hardy, this kohlrabi produces gorgeous violet-blue, frosted globes that mature in 70 days.
- **'EARLY PURPLE VIENNA' KOHLRABI:** Early maturing at 55 days from seeding, this cold hardy kohlrabi has delicious cabbage-flavored bulbs that have purple skin with white flesh.
- **'KOLIBRI' KOHLRABI:** This is a very cold hardy kohlrabi with purple skin and crisp, white flesh. Ready for harvesting in 45 days.

This vegetable is a nutritional powerhouse, providing a rich source of vitamin C, vitamin A, calcium, iron, and potassium.

RED NAPA CABBAGE

Red napa cabbage, also known as red Chinese cabbage, is a cherished vegetable that I cultivate from autumn to spring. Although it is a biennial plant, I cultivate it as an annual in my garden here in Southern California. It exhibits a diverse range of varieties and colors, including green, yellow, and pink. Red napa cabbage possesses a delightful combination of sweetness and crispness, making it suitable for various culinary applications. It can be consumed raw, steamed, incorporated into soups, and particularly excels in the preparation of kimchi.

Planting

Napa cabbage can be cultivated from autumn to spring in warmer climates. To avoid the summer heat, plant it early in spring. It thrives in temperatures between 50°F and 75°F (10°C–24°C) and is susceptible to rapid bolting when temperatures exceed 80°F (27°C). Red napa cabbage prefers full sun to partial shade, with sun for at least 6 hours daily. Like other Chinese cabbages and bok choy, it grows well in fertile and loose soil.

Sowing Indoors

Red napa cabbage can be propagated from seeds and is more likely to thrive when seedlings are started indoors and transplanted later in spring and fall. For spring planting, sow seeds 3 to 4 weeks before the last frost date. For fall planting, sow seeds 5 to 7 weeks before the first frost date.

Fill seeding trays with an organic seed-starting mix, plant seeds approximately ½ inch (1 cm) deep, and sow 1 to 2 seeds per cell. Maintain a consistent soil moisture during the germination period and place the seeding tray in a location with an air temperature between

60°F and 75°F (16°C–24°C). If necessary, use a heat mat to help seed germination. Remove the seeding tray from the heat mat once seeds have germinated and place it under grow lights to prevent leggy seedlings.

Transplants

Red napa cabbage seeds typically germinate within 3 to 5 days. The seedlings are ready for transplanting when they have developed a few sets of true leaves or have grown to approximately 3 inches (7.5 cm) in height. Transplant the seedlings into raised beds or in-ground gardens, spacing them 12 to 14 inches (30–36 cm) apart. If you're transplanting into containers, select a pot that is at least 10 inches (25 cm) wide and 12 inches (30 cm) deep.

Care

Water

Red napa cabbage requires consistent watering to thrive. If not, the leaves will become tough, bitter, and the plant will produce flowers and seeds prematurely. In raised beds, red napa cabbage needs approximately 1 inch (2.5 cm) of water per week. However, in containers, watering needs to be more frequent. The soil should never be allowed to dry out, but also should not become soggy.

Fertilizer

Red napa cabbage prefers fertilizers high in nitrogen to promote leaf growth. Fertilize the plants at the base once every 3 weeks.

Pests and Diseases

The most common pests of red napa cabbage are cabbage loopers, cabbage worms, aphids, slugs, and snails. (See page 150.)

Companion Plants

Red napa cabbage grow best with beets, celery, spinach, thyme, carrots, onions, bush beans, cucumbers, dill, potatoes, rosemary, and nasturtiums. Avoid growing red napa cabbage with radishes, kale, cabbages, cauliflowers, broccoli, and arugula.

Harvest and Storage

When to Harvest

Red napa cabbage cultivars fully mature within 50 to 70 days after transplanting. However, you can harvest the cabbage as soon as it develops a few sets of leaves.

For optimal flavor, red napa cabbage is best harvested in the morning. Red napa cabbage should be harvested as soon as it reaches maturity to prevent it from bolting. To harvest, cut the cabbage head above the ground and remove the outer leaves.

How to Store

After harvesting, refrigerate red napa cabbage in a sealed bag for up to 1 month. Alternatively, blanch the cabbage and freeze it for long-term storage. Kimchi and sauerkraut are also effective methods for preserving red napa cabbage.

My Favorite Vibrant Varieties

- **'HILTON' NAPA CABBAGE:** 'Hilton' has a medium-sized head and is mild-tasting and crunchy.
- **'MERLOT' NAPA CABBAGE:** This dense napa cabbage has a beautiful, vibrant pink color.
- **'ONE KILO SLOW BOLT' NAPA CABBAGE:** This cabbage has a large size with tight heads and a delicate flavor and softer texture.
- **'RED DRAGON' NAPA CABBAGE:** This variety's crisp, flavorful, oval leaves are green on the outside and dark purple on the inside.

↑ 'Red Dragon' napa cabbage seedlings beginning to develop their true leaves

PURPLE BOK CHOY

Bok choy, also known as pak choy or Chinese cabbage, is an ornamental plant belonging to the Brassicaceae family, which includes broccoli, cauliflower, and other similar vegetables. Bok choy exhibits a diverse range of sizes and colors, including purple. It is a widely recognized Chinese green that is used to make many delectable stir-fry dishes and soups. It is a rapidly growing plant characterized by crisp, sturdy stalks and tender leaves with a flavor reminiscent of Swiss chard. Bok choy is a nutrient-dense food source, providing an abundance of fiber, calcium, vitamin K, and vitamin C, as well as various other essential minerals.

Planting

Purple bok choy, like many other leafy greens, thrives during the cooler temperatures of the autumn to spring seasons. It flourishes in full sun, needing between 6 and 8 hours of sunlight daily—particularly during the fall season. However, in spring, bok choy tends to bolt rapidly if the conditions are too warm and sunny, necessitating growing them in partial shade with approximately 3 to 5 hours of sunlight per day. Bok choy prefers nutrient-rich and well-drained soil for optimal growth.

Sowing Indoors
Bok choy is an easy-to-grow crop that can be propagated from seed. Indoor seed starting is a more reliable method than starting seeds outdoors. This allows for the transplanting of seedlings into the garden later in the spring and fall.

In the spring, begin planting seeds 3 to 4 weeks before the anticipated last frost date. For fall planting, initiate seed planting 5 to 7 weeks before the average first frost date.

Fill your seeding trays with an organic seed-starting mix. Plant

It is a widely recognized Chinese green that is used to make many delectable stir-fry dishes and soups.

the bok choy seeds approximately ½ inch (1 cm) deep and with 2 to 3 seeds per cell. Maintain even soil moisture during the germination period. Place the seed tray in a location with a temperature between 60°F and 75°F (16°C–24°C). If necessary, use a heat mat to help faster and more even germination.

Remove the heat mat once the seeds have germinated and provide them with grow lights to prevent getting leggy seedlings.

Transplants

Bok choy seeds typically germinate within a 4 to 8 day period. The seedlings are ready for transplanting once they have developed a few sets of their true leaves. Plant bok choy seedlings in raised beds or directly in the ground, spacing them approximately 6 to 10 inches (15–25 cm) apart. If transplanting into containers, ensure that the container is at least 10 to 12 inches (25–30 cm) wide and deep.

Care

Water

Purple bok choy requires consistent watering to maintain its optimal growth. If not, the leaves will become tough and bitter, and the plant will bolt, producing flowers and seeds prematurely. In raised beds, bok choy needs approximately 1 inch (2.5 cm) of water per week. However, when grown in containers, watering needs to be more frequent.

Fertilizer

Bok choy prefers a high-nitrogen fertilizer to promote leaf growth. Fertilize at the base of the plants once every 3 weeks.

Pests and Diseases

Purple bok choy isn't prone to having problems with diseases. However, the most common pests for bok choy are cabbage loopers, cabbage worms, and slugs. (See page 150.)

↑ A harvest of beautiful purple bok choy

Companion Plants

Purple bok choy grow best with beets, celery, thyme, carrots, onions, bush beans, cucumbers, dill, potatoes, rosemary, and nasturtiums. Avoid growing bok choy with radishes, kale, cabbages, cauliflowers, broccoli, and arugula.

Harvest and Storage

When to Harvest

Bok choy cultivars fully mature in 50 to 75 days, but you can harvest them as soon as they have a few sets of leaves. Smaller types mature at about 6 to 8 inches (15–20 cm) tall and the larger varieties can grow up to 12 to 18 inches (30–46 cm) tall. Harvest bok choy in the early morning after the dew dries when they are at their sweetest and crispiest.

Bok choy is the type of plant that is a cut-and-come-again, so there are a couple ways to harvest bok choy. First, you can harvest from the outermost layer of leaves, leaving the new leaves to grow from the center of the plant. Second, you can cut the entire plant, leaving about 2 inches (5 cm) of stem in the ground and the bok choy plant will regrow. Lastly, just harvest the whole plant.

How to Store

After harvesting, do not wash bok choy; simply store the leaves in a clear, plastic, sealable bag in the refrigerator for up to 1 week. You can also blanch and freeze it for long-term storage.

My Favorite Vibrant Varieties

- **'BABY MILK' BOK CHOY:** A bok choy with cream-colored stems and a green leaf, this bok choy has a sweet taste and crunchy texture.
- **'PURPLE LADY' BOK CHOY:** A beautiful eye-catching purple leaf bok choy, it's sweet and rich in flavor.
- **'SUZHOU' BOK CHOY:** This medium-sized bok choy has a great flavor and is the most popular bok choy at the market.
- **'YELLOW HEART WINTER' BOK CHOY:** The plant forms a beautiful rosette; it is cold hardy and sweet with extra-tender leaves.

RAINBOW SWISS CHARD

Rainbow Swiss chard is a vibrant ornamental vegetable commonly cultivated in gardens. It exhibits a diverse array of colors and varieties. Although belonging to the beet family, Swiss chard does not produce edible roots. Its leaves are the edible part, and when cooked, the flavor closely resembles that of beet leaves. Rainbow Swiss chard is a low-calorie food that provides an abundance of fiber, vitamin A, vitamin C, and iron.

Planting

Swiss chard is an exceptionally hardy and straightforward vegetable to cultivate, making it an ideal choice for beginners to grow. Rainbow Swiss chard thrives in cool weather during spring and fall, although it can tolerate moderate heat. All types of Swiss chards prefer loose, nutrient-rich, and well-drained soil. While it prefers full sun for at least 6 hours per day, Swiss chard can tolerate partial shade during the summer months in warmer climates.

Direct-Sowing
To help the germination process, rainbow Swiss chard seeds can be soaked in warm water overnight. Swiss chard can be direct-sown after the last frost or as soon as the soil is workable, and when the soil temperature reaches between 55°F to 75°F (13°C–24°C). Sow the seeds ½ inch (1 cm) deep and approximately 8 to 10 inches (20–25 cm) apart. Typically, the seeds will germinate within 1 week.

Sowing Indoors
Sow seeds indoors 2 to 4 weeks before the final frost date. Fill the seedling tray with seed-

↑ Swiss chard makes a bold statement in the garden.

Its leaves are the edible part, and when cooked, the flavor closely resembles that of beet leaves.

starting mix and sow 2 seeds per cell, ensuring they are planted ½ inch (1 cm) deep. Maintain even soil moisture during the germination period and position the seedling tray at room temperature between 60°F and 75°F (16°C–24°C). If necessary, use a heat mat for seed germination. Remove the heat mat once the seeds have germinated and place the tray under a grow light to prevent leggy seedlings. Chard seedlings are ready for transplanting when they have developed a few sets of their true leaves.

Container Planting

Swiss chard can also be grown in containers. The container should be at least 12 inches (30 cm) wide and 12 to 14 inches (30–36 cm) deep.

Care

Water

Swiss chard requires regular watering to promote optimal leafy growth and overall health—despite its relatively drought-tolerant nature once it is mature. If you are growing rainbow Swiss chard in a container, water it 2 to 3 times per week throughout the growing season. Additionally, applying mulch can help with water retention.

Thinning

If you direct-sow rainbow Swiss chard seeds outdoors, thin out the seedlings once they have grown to 2 to 3 inches (5–7.5 cm) tall so they are spaced 4 to 8 inches (10–20 cm) apart.

Fertilizer

Rainbow Swiss chard prefers nitrogen-based fertilizers to promote good leaf growth. Apply the fertilizer at the base of the plants every 3 weeks.

Pests and Diseases

Rainbow Swiss chard tend to not have a lot of problems with pests or diseases, because the plants grow fast and are more likely to be harvested often. However, the most common pest

↑ Gorgeous rainbow Swiss chard and its colorful root system

and disease for Swiss chard are aphids and powdery mildew. (See page 150.)

Companion Plants

Rainbow Swiss chard is in the same family as beets (Chenopodiaceae), so it grows well with garlic, carrots, kale, Brussels sprouts, mint, radishes, bush beans, lettuce, cabbage, broccoli, sage, and marigolds. Avoid growing rainbow Swiss chard with spinach and pole beans.

Harvest and Storage

Baby rainbow Swiss chard leaves are sweet and tender and can be harvested once they reach 5 to 6 inches (13–15 cm) in height. However, it is important to avoid over-harvesting, as this can hinder the plant's growth and development. Unlike some other vegetable plants, Swiss chard grows from the center of the plant, and the best way to harvest it is to remove the outer leaves first. By doing so, you encourage the plant to continue growing and prevent it from going to flower prematurely.

How to Store

Rainbow Swiss chard is most flavorful when consumed immediately after harvest. However, it can be stored in the refrigerator for up to 3 weeks. To preserve rainbow Swiss chard leaves, select only the healthy ones, remove them from the stalks, and store them separately from the stalks in a sealed, clear, plastic bag.

Another method for long-term storage is freezing.

My Favorite Vibrant Varieties

- **'BRIGHT LIGHTS' SWISS CHARD:** Glossy green or bronze leaves with stems of gold, pink, orange, red, or white; with a sweet, mild flavor, every color tastes a bit different.
- **'FORDHOOK GIANT' SWISS CHARD:** A giant chard that produces large green leaves with white stems.
- **RAINBOW SWISS CHARD OR FIVE-COLOR SILVERBEET SWISS CHARD:** Vibrant colored stems in shades of bright pink, gold, orange, yellow, white, and red.
- **'VULCAN' SWISS CHARD:** Leaves are a bright red color with a great flavor.

PINK CELERY

Pink celery boasts vibrant and aesthetically pleasing stalks. In Chinese culture, pink celery holds a symbolic significance, representing prosperity. It possesses an aromatic herbal scent and a robust, crunchy flavor with a touch of sweetness. Pink celery can be consumed raw, incorporated into salads, soups, and stews, or put in a stir-fry. Pink celery is a rich source of vitamins A, C, and K, as well as providing fiber and antioxidants. Additionally, its low-calorie content makes it a valuable addition to a balanced diet.

Planting

Celery is a cool-season crop that is not frost-tolerant. It can be propagated from seeds or kitchen scraps, or purchased as a transplant from nurseries. Celery thrives in a sunny location that receives 5 to 7 hours of sunlight daily. It prefers fertile, well-drained soil with ample organic matter for optimal growth. In regions with cooler spring and summer temperatures, plant celery in early spring for a summer harvest. Conversely, in areas with warmer springs and summers, plant celery in the late summer for a fall or winter harvest.

↑ 'Chinese Pink' celery

Sowing Indoors

Celery can be initiated indoors approximately 2 to 4 weeks before outdoor planting. First, fill the seeding trays with an organic seed-starting mix. Plant the seeds approximately ¼ inch (0.6 cm) deep and 1 to 2 seeds per cell. Maintain a consistent soil moisture during the germination period and use a heat mat to ensure temperatures remain between 60°F and 75°F (16°C–24°C). Germination typically occurs within 2 to 3 weeks.

Celery seedlings are ready for outdoor transplanting when they have reached a height of approximately 4 inches (10 cm) and have developed 5 to 6 leaves. Transplant the seedlings into the ground or into raised beds, spacing them 8 to 10 inches (20–25 cm) apart.

Growing from Kitchen Scraps

Regrowing kitchen scraps is a rewarding and enjoyable gardening activity. Celery is an excellent choice for this project, as it can be regrown from the base of the harvested stalks. There are two main methods for regrowing celery: placing the base of the celery in a bowl of water or planting it in soil.

In Water

Simply cut the celery stalks approximately 2 to 3 inches (5–7.5 cm) from the base and place the stalks in a small bowl filled with about 1 to 2 inches (2.5–5 cm) of water. Position the bowl in a bright indoor area. Within a few days, the celery stalks should begin to sprout roots. Remember to change the water every 1 to 2 days and ensure the bowl has water in it at all times.

In Potting Soil

If you desire, you can propagate pink celery indoors after harvesting the stalks from the garden. To do this, replant the harvested plant. Fill a pot with potting mix and after digging up the plant and harvesting the top, gently press the celery base about 1 inch (2.5 cm) into the soil. Maintain consistent soil moisture. Place the pot in a bright location and cover with a clear plastic bag if needed. Soon, new shoots should appear.

Care

Water

Pink celery is composed of 95% water, necessitating consistent watering. It requires at least 1 to 2 inches (2.5–5 cm) of water per week. Regular watering throughout the growing season is crucial to prevent stringy and tough stalks. Overhead watering should be avoided as it can lead to moisture accumulation in the leaves and stalks, potentially attracting pests and fungi. Mulch can help by retaining moisture in the soil and suppressing weed growth.

Fertilizer

Celery is a heavy feeder. Fertilize every 2 to 3 weeks with an all-purpose fertilizer, liquid fish fertilizer, compost tea, or another organic fertilizer.

Blanching

What is celery blanching? Blanching is a technique used to protect celery stalks from direct sunlight. Not all celery varieties require blanching; some are naturally self-blanching. This process enhances the paleness and sweetness of the stalks. After the stalks have grown to approximately 6 to 8 inches (15–20 cm) in height, row covers can be used to create shade and a cooler environment. Alternatively, individual plants can be blanched by mounding soil up in stages around the stalks or wrapping the stalks in newspaper or cardboard, blocking out the sun.

Pests and Diseases

The most common pest and disease for pink celery are aphids and powdery mildew. (See page 150.)

Companion Plants

Pink celery grows best with beans, chickpeas, mint, sage, oregano, dill, rosemary, thyme, garlic, onions, chives, shallots, leeks, cosmos, marigolds, nasturtiums, and brassicas. Avoid planting pink celery with corn, potatoes, carrots, and parsley.

Harvest and Storage

When to Harvest

Pink celery is ready for harvest when it reaches approximately 12 to 14 inches (30–36 cm) in height. You can harvest individual stalks or the entire plant once it is mature. To encourage continuous growth, harvest the outer layer of stalks as needed. This practice allows the celery plant to produce new stalks throughout the season.

How to Store

Pink celery stalks can be stored in the refrigerator to maintain their crispness. For extended shelf life, consider freezing them. Frozen pink celery is versatile in various culinary applications, such as soups, stews, and casseroles.

How to Freeze

1. **Preparation:** Thoroughly clean the celery stalks. Trim the leaves and dice or cut the celery stalks to your desired length.

2. **Blanching:** Blanching is optional, but can enhance celery flavor retention. Briefly boil the diced celery pieces for a few minutes and then immediately transfer them to an ice bath for a few minutes.

↑ Pink celery in the garden

3. **Drain and dry:** Drain the blanched celery and pat it dry with a clean cloth.

4. **Freeze:** Line a shallow tray with wax paper. Arrange the celery pieces in a single layer and freeze overnight. Subsequently, transfer the frozen celery into sealable, clear, plastic bags for long-term storage, up to 12 to 15 months.

My Favorite Vibrant Varieties

- **'CHINESE PINK' CELERY:** A stunning bubblegum pink celery from China, with a crunchy texture and lightly sweet taste.
- **'GIANT RED' CELERY:** Large stalks are a striking purplish-red and green color that turn pink when they are cooked; good cold hardiness.
- **'GOLDEN BOY' CELERY:** A unique celery with stalks that easily blanch to a golden-yellow color; both the stalks and the celery hearts are tender.

Leaf Vegetables

PURPLE CABBAGE

Purple cabbage belongs to the Brassicaceae family, which also includes broccoli, cauliflower, Brussels sprouts, collards, and kohlrabi. Purple cabbages are known for their hardiness, flavor, and nutritional value. They are rich in vitamin A, vitamin B6, and various antioxidants. Additionally, they contain twice the amount of iron and vitamin C compared to green cabbage. Purple cabbage can be consumed raw or prepared in various dishes, such as coleslaw, sauerkraut, stir-fries, stews, and soups.

Planting

Growing purple cabbage is comparable to cultivating green cabbage. Purple cabbage is a cool-season vegetable that thrives in sunny, partially shaded locations. It prefers well-drained soil with a high organic matter content for optimal growth. It can be propagated from seed or transplants.

Sowing Indoors

To ensure a spring harvest of winter cabbages, sow cabbage seeds indoors approximately 6 to 8 weeks before the anticipated last frost date in the spring. Fill the seedling tray with a seed-starting mix and sow 1 to 3 seeds per cell, ensuring they are planted ¼ to ½ inch (0.6–1 cm) deep. Maintain consistent moisture during the germination period and position the seedling tray in a location at room temperature between 60°F and 70°F (16°C–21°C). Alternatively, use a heat mat to help seed germination. Once the seeds have germinated, remove the heat mat and provide them with indirect sunlight to prevent leggy seedlings. Purple cabbage seedlings are ready for transplanting when they have developed a couple of sets of their true leaves.

Transplants

Purple cabbage seeds typically germinate within a 7-to-14-day period. The seedlings are ready for transplanting once they have developed a few sets of their true leaves, which usually takes approximately 5 to 8 weeks. Plant purple cabbage seedlings in raised beds or directly in the ground, spacing them 18 to 24 inches (46–61 cm) apart. If transplanting into containers, ensure that the container is at least 10 to 12 inches (25–30 cm) wide and deep.

Care

Water

Cabbage requires consistent watering, approximately 1 to 2 inches (2.5–5 cm) per week. Applying mulch around the plants aids in water retention and suppresses weed growth. Inconsistent watering during the heading stage can result in split cabbages.

Fertilizer

Cabbage is a heavy feeding crop. Apply nitrogen-based fertilizer for a few weeks after transplanting to stimulate plant growth. Then refrain from applying any additional nitrogen once the head begins to develop. Excessive nitrogen at this stage can result in loose heads and splitting.

Row Covers

Protect cabbage plants from pests such as slugs, snails, cabbage worm moths, and aphids by covering the entire crop with mesh or row cover fabric. Use PVC pipe hoops as the supporting structure for the row covers. Row covers also will deter birds from damaging your crop.

Pests and Diseases

The most common pests and diseases of purple cabbage are aphids, cabbage worms, cabbage looper, and leaf spot. (See page 150.)

↑ Purple cabbage seedlings ready to be transplanted into a raised bed

Companion Plants

Purple cabbage and other brassicas grow well with chamomile, rosemary, sage, mint, and dill. Avoid growing cabbage with eggplants, peppers, tomatoes, and potatoes.

Harvest and Storage

When to Harvest

There are several indicators to determine when purple cabbages are ready for harvesting. Firstly, knowing the days to harvest period is crucial. Most cabbages mature within 60 to 80 days, while some varieties with large and dense heads can take up to 180 days. Secondly, purple cabbage is ready for harvest when the head becomes firm to the touch. If left unharvested for an extended period when mature, the head may split.

The optimal time to harvest purple cabbages is in the morning, as they retain moisture overnight, resulting in crispier and sweeter heads. Purple cabbages can be harvested when they reach the desired size. Use a sharp, clean garden knife to cut close to the base of the cabbage head. Avoid breaking or snapping the cabbage head off the stem, as this can damage both the head and the plant, if you intend for the stem to regrow another head. After the last harvest, remove the plant entirely.

How to Store

Purple cabbages are at their peak flavor when enjoyed fresh after harvest. You can also store the cabbage head in sealed, clear, plastic bags and refrigerate it. Do not wash the cabbage before refrigerating.

You can preserve your cabbage by making sauerkraut, and I particularly enjoy making pickled cabbage.

My Favorite Vibrant Varieties

- **'GLORY OF ENKHUIZEN' CABBAGE:** A green cabbage that originated in Holland; beautifully round with a medium to large head.
- **'RUBY PERFECTION' CABBAGE:** A mid-size red cabbage that produces dense heads between 4 and 6 pounds (2–3 kg).
- **'SAPPORO GIANT' CABBAGE:** A large, green cabbage from Japan that can reach 44 pounds (20 kg) per head.
- **'TETE NOIRE' CABBAGE:** From France, a red cabbage with smaller, deep red heads.
- **'TIARA' CABBAGE:** A green cabbage with smaller heads of between 1 and 2 pounds (0.5–2 kg); delicious and mildly sweet.
- **'TROPIC GIANT' CABBAGE:** A green cabbage that is heat tolerant and disease resistant; perfect for southern gardeners.

PINK-VEINED COLLARD GREENS

Pink-veined collard greens are members of the brassicas, along with broccoli, cauliflower, cabbage, kale, and Brussels sprouts. Collards are another leafy vegetable that is highly nutritional, very easy to cultivate, and is cold hardy.

Planting

Collards are cool-season vegetables commonly planted in late summer or early fall to ensure a winter and spring harvest. They thrive in full sun or partial shade sites that receive approximately 6 hours of sunlight daily. Pink-veined collard greens prefer well-drained and fertile soil with a pH level between 6.5 and 6.8. They can be grown from seeds indoors and transplanted outdoors into the garden.

Sowing Indoors

For spring planting, sow collard seeds 4 to 6 weeks before the anticipated last frost date. For fall planting, sow collard seeds midsummer approximately 2 months before the anticipated first frost date.

Fill a seedling tray with a seed-starting mix and sow 1 to 3 seeds per cell, ensuring the seeds are planted ¼ to ½ inch (0.6–1 cm) deep. Maintain even soil moisture during the germination period and place the seedling tray in a location with air temperatures between 60°F and 70°F (16°C–21°C). If necessary, use a heat mat to help seed germination. Collard seeds are characterized by their rapid germination, typically occurring within 5 to 7 days. Remove the heat mat once the seeds have germinated and position the tray under a grow light to prevent leggy seedlings.

Transplants

Once the seedlings have developed a few sets of true leaves, they are ready for transplanting. Plant collard seedlings outside in raised beds or directly in the ground, spacing them 18 to 24 inches (46–61 cm) apart. If transplanting in containers, select a pot with a width of 10 to 12 inches (25–30 cm) and a depth of at least 12 inches (30 cm).

Care

Water

Collard greens require consistent watering to maintain optimal growth. Ensure that the soil remains consistently moist but not waterlogged. Approximately 1 to 2 inches (2.5–5 cm) of water per week is necessary. Watering should be directed at the base of the plants, but avoid wetting the foliage. This practice helps prevent fungal diseases like powdery mildew.

To enhance soil moisture retention, suppress weeds, and regulate soil temperature, apply organic mulch around the plants.

Pink-veined collards grow well with beans, beets, celery, herbs, onions, and potatoes.

↑ Pink-veined collards in the garden

Fertilizer

Before planting, apply organic compost and a slow-release, all-purpose fertilizer to the collard greens. During the growing season, apply fish fertilizer every 3 to 4 weeks to promote vigorous growth.

Pests and Diseases

The most common pests and diseases of pink-veined collards are aphids, cabbage worms, flea beetles, and powdery mildew. (See page 150.)

Companion Plants

Pink-veined collards grow well with beans, beets, celery, herbs, onions, and potatoes. Avoid growing collards with other brassicas like cabbage, cauliflower, broccoli, as they attract the same pests.

Harvest and Storage

When to Harvest

Pink-veined collards can be harvested at various stages of development. Baby collards, harvested when the leaves are approximately 2 to 5 inches (5–13 cm) long, offer tender and mild-flavored leaves, as more mature leaves can become tough and bitter. Alternatively, individual leaves can be harvested when they reach a width and length of 5 to 6 inches (13–15 cm). Commence harvesting from the outermost layers to allow the center leaves to continue growing and produce more for later harvests. Lastly, the entire plant can be cut off at the base for a larger harvest. I typically perform this method at the end of the season to clear space for next season's crops.

How to Store

Pink-veined collard greens are best consumed fresh, but they can also be stored for later use when you can add them to soups, stews, or stir-fries. Collard greens can be stored in plastic bags without washing and kept in the refrigerator for up to 1 week. Alternatively, they can be blanched and stored in the freezer.

My Favorite Vibrant Varieties

- **'CHAMPION' COLLARDS:** A collard that produces high yields, it can grow from 24 to 36 inches (61–91 cm) and spread about 30 inches (77 cm).
- **'GEORGIA' COLLARDS, 'GEORGIA SOUTHERN' COLLARDS, OR 'GEORGIA HYBRID' COLLARDS:** This tall collard with smooth leaves and white or pale green stalks can grow up to 6 feet (183 cm) tall.
- **'OLE TIMEY BLUE' COLLARDS:** A unique cultivar with added ornamental appeal, it is pink-veined with blue-green leaves, growing up to 2 feet (61 cm) tall. It has very good eating qualities too.
- **'TIGER HYBRID' COLLARDS:** A high-yielding cultivar with large, thick, and slightly savoyed leaves, it is known for having a sweet flavor. The mature height is from 20 to 24 inches (51–61 cm) with a spread of 24 inches (61 cm).

PURPLE KALE

Purple kale is a biennial and member of the Brassicaceae family, which contains many similar plants such as cabbage, broccoli, cauliflower, and Brussels sprouts. Purple kale is a hardy and resilient plant that can thrive in various climates. The edible kales can also be grown as an ornamental plant, featuring diverse leaf colors like white, pink, and purple, along with textured and curly leaves. Purple kale is a rapid-growing vegetable that is very productive and is exceptionally easy to cultivate; it is a versatile ingredient for cooks, making it a valuable addition for any diet. Kale is widely recognized as a superfood due to its high antioxidant content. It can be consumed raw or incorporated into various cooked dishes, including soups and sautéed meals. It has gained popularity in recent years as an ingredient in smoothies and juices. And then there's the benefits of eating crispy, flavor-packed kale chips!

Planting

Purple kale thrives in cool-weather conditions, but it can also tolerate moderate temperatures. Kale can be grown either indoors or sown directly into the garden outdoors.

Sowing Indoors

Fill the seedling tray with seed-starting mix and sow 1 to 3 seeds per cell, ensuring they are planted ¼ to ½ inch (0.6–1 cm) deep. Maintain even soil moisture during the germination period. Place the seedling tray in a location that has a temperature of between 55°F and 70°F (13°C–21°C). If necessary, use a heat mat to help seed germination. Remove the heat mat once the seeds have germinated and position the tray under a grow light to prevent leggy seedlings.

Transplants

Purple kale seeds will exhibit rapid germination, typically germinating within 3 to 5 days. The seedlings are ready for transplanting when they have devel-

↑ Even the stems are colorful on purple kale!

Purple kale is a rapid-growing vegetable that is very productive and is exceptionally easy to cultivate; it is a versatile ingredient for cooks, making it a valuable addition for any diet.

oped a few sets of true leaves. Plant seedlings in raised beds or in-ground gardens, spacing them 18 to 24 inches (46–61 cm) apart. If transplanting the seedlings in containers, select a pot with a width of 10 to 12 inches (25–30 cm) and a depth of at least 12 inches (30 cm).

Direct-Sowing

Sow kale seeds outdoors as soon as the ground is workable; typically this is 4 weeks before the last frost date in the spring. Alternatively, you can plant kale seeds outdoors in late August or early September, 4 to 6 weeks before the first frost date, and enjoy harvesting it in the autumn. Kale prefers full sun or partial shade in locations that receive at least 6 hours of sunlight per day. Kale thrives in fertile, well-drained soil with a pH between 6 and 7.5.

Care

Water

Consistent moisture is paramount for kale's good health, particularly during dry periods. Purple kale requires approximately 1 to 1½ inches (2.5–4 cm) of water per week. To avoid overwatering, use soaker hoses or irrigation systems to maintain dry foliage, thereby preventing leaf diseases.

Apply organic mulch around kale plants to enhance soil moisture retention, suppress weeds, and regulate soil temperature.

Fertilizer

Before planting, apply organic compost and a slow-release, all-purpose fertilizer to the purple kale site. During the growing season, apply fish fertilizer every 3 to 4 weeks to support vigorous plant growth.

Pests and Diseases

The most common pests and diseases of kale are cabbage worms, aphids, flea beetles, slugs, snails, and powdery mildew. (See page 150.)

Companion Plants

Purple kale grows well with beans, beets, celery, herbs, onions, and potatoes. Avoid growing kale with other brassicas like cabbage, cauliflower, and broccoli, as they attract the same pests.

Harvest and Storage

When to Harvest

Purple kale can be harvested at various stages. Baby kale can be harvested when the leaves are approximately 2 to 5 inches (5–13 cm) long. Harvesting early results in tender and mild-flavored leaves. Alternatively, individual leaves can be harvested when they are about 5 to 6 inches (13–15 cm) wide and 5 to 10 inches (13–25 cm) long, starting with the outermost layer to allow the central leaves to continue growing. Lastly, the entire plant can be cut off at the base for a larger harvest. I typically do this at the end of the season to clear space for next season's crops.

How to Store

Purple kale is most suitable for consumption fresh, but it can

↑ A delicious harvest, ready for the kitchen

also be preserved for later use. Kale can be stored fresh in plastic bags without washing the leaves and kept in the refrigerator for up to 1 week. I prefer, though, to store kale in the freezer. First, I chop the kale leaves and place them in plastic freezer bags. These can then be stored in the freezer for a few months, then thawed and added to soups, stews, and smoothies.

My Favorite Vibrant Varieties

- **CURLY KALE:** This kale has ruffled leaves with a pungent flavor and is one of the most common types found in grocery stores.
- **LACINATO KALE, DINOSAUR KALE, OR TUSCAN KALE:** This kale is originally from Italy and has long, dark green, bumpy leaves with a slightly sweet taste.
- **'RED RUSSIAN' KALE:** This kale has flat, fringed leaves and purple veins; it has a tender texture with a mild flavor.

MULTICOLORED LETTUCE

Multicolored lettuce is a highly rewarding vegetable to cultivate, requiring minimal effort and resulting in substantial cost savings. Its rapid growth, high yield, and low maintenance make it an ideal choice for gardeners. Homegrown lettuce possesses a superior flavor compared to store-bought varieties, and it is an excellent source of essential vitamins and minerals.

Planting

Multicolored lettuce, along with many other attractive lettuce types, are a cool-weather crop that can withstand light frosts. They thrive in sunny locations, preferring partial shade to full sun. Lettuce requires fertile, well-drained, and nutrient-rich soil. In warmer regions with a warm winter, lettuce can be grown from fall to spring. Its versatility allows it to be grown in containers, vertical gardens, and balconies—making it ideal for small spaces.

Residing in Southern California's coastal region, I have the opportunity to cultivate lettuce year-round (with appropriate protection during the summer months). Multicolored lettuce can be direct-sown, started indoors from seed, or transplanted from nursery plants.

Direct-Sowing

Direct-sow multicolored lettuce seeds in the early spring, 4 weeks before the last frost date in your region. Sow lettuce seeds 4 to 8 inches (10–20 cm) apart in a row and 10 to 12 inches (25–30 cm) apart between the rows. Plant lettuce seeds ¼ inch (0.6 cm) deep to ensure adequate light for germination. Alternatively, scatter lettuce seeds

COMMON LETTUCE VARIETIES

There are two primary types of lettuce: **loose leaf** and **head** lettuce. Within each of these two types there is a diverse array of varieties, colors, and leaf shapes for gardeners to try.

- **Butterhead Type:** One of the most popular lettuce varieties is butterhead, characterized by its crisp and fresh flavors. 'Bibb' and 'Boston' are the most popular varieties of the butterhead lettuce group. They have a cuplike shape and are characterized by their sweet, tender, and buttery texture.
- **Crisphead Type:** The most popular variety of the crisphead type is 'Iceberg' lettuce, characterized by its compact head and crisp, light green leaves.
- **Romaine Type:** Romaine lettuce types have elongated, thick leaves that are crisp and sturdy with firm ribs. Romaine is quite heat tolerant, unlike many other types. Each leaf is shaped like a canoe and grows upright on a tight head.
- **'Little Gem' Lettuce:** This cross between a romaine and butterhead lettuce has a sweet flavor and crisp texture. 'Little Gem' resembles a romaine lettuce type, but is smaller.
- **Leaf Lettuce Type:** This lettuce type is fast growing and long lasting, with green or red leaves.
- **Oak Leaf Type:** The name for this type is derived from the distinctive leaf shape that resembles the leaves of oak trees. Oak leaf varieties can be found in either green or red colors.

lightly over the planting area without burying them. Gently dust some soil on top to provide support for the seeds. Water thoroughly and maintain even soil moisture. After germination, reduce the watering frequency.

Sowing Indoors

Fill the seedling tray with seed-starting mix and sow 2 to 4 seeds per cell, ensuring they are planted ¼ inch (0.6 cm) deep. Maintain even soil moisture during the germination period.

Place the seedling tray in a location with an air temperature from 55°F to 65°F (13°C–18°C). If necessary, use a heat mat for seed germination. Remove the heat mat once the seeds have germinated and position the tray under a grow light. Lettuce seeds typically germinate within approximately 1 week.

Transplants

Multicolored lettuce seedlings can be transplanted when they reach a height of 3 to 4 inches (7.5–10 cm) or have developed at least two sets of true leaves. Before transplanting, it is essential to acclimate the seedlings to outdoor conditions for 1 week. This process involves gradually exposing the seedlings to outdoor light, winds, and temperatures for increasing periods each day.

When transplanting, space the lettuce seedlings approximately 8 to 10 inches (20–25 cm) apart in the row and the rows 12 inches (30 cm) apart. It is recommended to transplant in the late afternoon or evening to allow the seedlings to acclimate to the sun without experiencing excessive heat stress.

Succession Planting

Succession planting is a method of gardening that involves sowing seeds every 1 to 2 weeks. This technique extends the harvest period, ensuring a steady supply of vegetables throughout the growing season.

Successful for most cool-weather crops—including lettuce, carrots, spinach, radishes, peas, Swiss chard, and bok choy, among others—succession

Multicolored lettuce is a highly rewarding vegetable to cultivate, requiring minimal effort and resulting in substantial cost savings.

↑ Lettuce seedlings ready to be transplanted into a raised bed

planting provides a reliable harvest without the challenges of overcrowding or a sudden abundance of produce.

Care

Water

Multicolored lettuce plants, like many other lettuce types, have a shallow root system and require frequent watering. The optimal time to water lettuce is in the morning, and it is crucial to prevent the soil from drying out. Therefore, it is recommended to deeply water lettuce once a week. Applying an organic mulch around the base of lettuce plants can also help in retaining soil moisture, suppressing weeds, and regulating soil temperature.

Fertilizer

Before planting, apply organic compost and a slow-release, all-purpose fertilizer to the soil. During the growing season, apply fish fertilizer (fish emulsion) every 3 weeks to support leaf growth.

Pests and Diseases

The most common pests and diseases for lettuce are aphids, slugs, snails, caterpillars, and powdery mildew. (See page 150.)

Companion Plants

Lettuce grows well with beets, carrots, asparagus, calendula, cilantro, eggplants, chives, garlic, mints, radishes, nasturtiums, onions, turnips, and parsnips. Avoid growing lettuce with vegetables in the Brassicaceae family and fennel.

Harvest and Storage

When to Harvest

Multicolored lettuce can be harvested at various stages of development. Head lettuce types typically mature in approximately 6 to 8 weeks after transplanting, while loose leaf varieties are ready in just 4 weeks. There are three primary methods for harvesting lettuce:

- **Cutting the outermost leaves:** This method allows new leaves to grow from the center of the plant.
- **Cutting the plant at soil level:** Leave approximately 2 inches (5 cm) of stem above the soil to allow the lettuce to regrow.
- **Harvesting the entire plant:** Harvest the entire mature plant.

How to Store

Multicolored lettuce is most flavorful when consumed fresh. Therefore, it is advisable to harvest only what you intend to use immediately. However, if you have an excess harvest, proper storage can extend its shelf life to up to 7 to 10 days. To maintain freshness, avoid washing the lettuce and store it in a plastic bag in the refrigerator. When ready to use, simply rinse it with cold water and soak it for a few minutes.

My Favorite Vibrant Varieties

- **'BLACK SEEDED SIMPSON' LETTUCE:** This leaf lettuce that has green, ruffled leaves. It is one of the most tender and flavorful lettuces, and is adapted to a range of climates.
- **'BUTTERCRUNCH' LETTUCE:** Buttercrunch has a luscious and buttery texture. It is a butterhead type with a soft head, and some heat resistance.
- **GIANT ROMAINE LETTUCE:** This is a larger romaine head than normal varieties. It has a crisp texture, with a dense head, and can be stored in the fridge for up to 2 weeks.
- **'ITHACA' LETTUCE:** This improved 'Iceberg' has a firm, well-wrapped head that can grow to about 5 to 6 inches (13–15 cm) across with a crisp texture.
- **'LITTLE GEM' LETTUCE:** This romaine type has small, robust green heads. It's like growing just the heart of a romaine lettuce.

PURPLE MUSTARD GREENS

Purple mustard greens are a striking addition to the fall-planted vegetable garden. 'Osaka Purple' mustard greens are one of the best. They are characterized by deep violet-purplish and green veins and stems. This vibrant plant can reach a height of up to 24 inches (61 cm) in my garden, creating an impressive bouquet-like appearance. Not only is 'Osaka Purple' mustard visually appealing, but it also boasts a unique flavor and is highly productive.

Planting

Purple mustard, similar to green mustard, is a cool-season vegetable that thrives in sunny locations with fertile and well-drained soils. It can be grown from seed or transplants. The optimal time for cultivation is during the cooler months of early spring or late summer until early fall. Purple mustard prefers a slightly acidic to neutral soil pH range of between 6.0 and 7.5.

Direct-Sowing

Mustard thrives in cool weather and can tolerate light frosts. In spring, direct-sow seeds outdoors 4 to 6 weeks before the last frost date in your area. For a fall harvest, plant seeds outdoors approximately 6 to 8 weeks before the first frost date. Succession planting of purple mustard provides a continuous harvest of tender, flavorful mustard leafy greens throughout the cooler season.

Sow purple mustard seeds about ¼ inch (0.6 cm) deep and 8 to 12 inches (20–30 cm) apart. After sowing, lightly cover the seeds with soil and maintain a consistent moisture during the germination period, which typically occurs within 6 to 10 days.

Sowing Indoors

For spring planting, sow mustard seeds inside 4 to 6 weeks before the anticipated last frost date. For fall planting, sow mustard seeds in midsummer approximately 6 to 8 weeks before the anticipated first frost date.

Fill the seedling tray with a seed-starting mix and sow 1 to 3 seeds per cell, ensuring a depth of ¼ to ½ inch (0.6–1 cm). Maintain even soil moisture during the germination period. Purple mustard seeds exhibit rapid germination, typically requiring approximately 5 to 10 days. Additionally, place the seedling tray in a location with air temperatures between 60°F and 70°F (16°C–21°C). If necessary, use a heat mat for seed germination. Remove the heat mat once the seeds have germinated and position the tray under a grow light to prevent leggy seedlings.

It boasts a unique flavor and is highly productive.

Transplants

The mustard seedlings are ready for transplanting when they have developed a few sets of true leaves. Transplant purple mustard seedlings into raised beds or into in-ground gardens, spacing them 10 to 12 inches (25–30 cm) apart. If transplanting into containers, select a pot with a width of 10 to 12 inches (25–30 cm) and a depth of at least 12 inches (30 cm).

Care

Water

Purple mustard requires consistent watering. Ensure the soil remains consistently moist but not waterlogged. Approximately 1 inch (2.5 cm) of water per week is necessary. Watering should be directed at the base of the plants, avoiding wetting the foliage. To enhance soil moisture retention, suppress weeds, and regulate soil temperature, apply an organic mulch around the plants.

Fertilizer

Before planting, apply organic compost and a slow-release, all-purpose fertilizer to the soil. Additionally, apply fish fertilizer every 3 to 4 weeks during the growing season to support vigorous growth.

Pests and Diseases

The most common pests and diseases of purple mustard are aphids, cabbage worms, flea beetles, and powdery mildew. (See page 150.)

Companion Plants

Purple mustard greens grow well with beans, beets, celery, herbs, onions, and potatoes. Avoid growing purple mustards with other brassicas like cabbage, cauliflower, and broccoli, as they attract the same pests.

Harvest and Storage

When to Harvest

Purple mustards can be harvested at various stages. Baby mustards can be harvested when the leaves are approximately 2 to 5 inches (5–13 cm) long. The tender and mild-flavored young leaves are preferred over mature leaves, which can become tough and bitter. Alternatively, individual leaves can be harvested when they reach a width of 5 to 6 inches (13–15 cm) and a length of 5 to 10 inches (13–25 cm). This method involves starting with the outermost layer to allow the central leaves to continue growing. Lastly, the entire plant can be cut at the base for a larger harvest. I typically perform this method at the end of the season to clear space for next season's crops.

How to Store

Purple mustard greens are the most flavorful when consumed fresh. However, they can also be stored for later use, such as adding them to soups, stews, and stir-fries. Purple mustard leaves can be stored in plastic bags without washing and kept in the refrigerator for up to a week.

↑ Frilly leaf purple mustard

My Favorite Vibrant Varieties

- **DRAGON TONGUE MUSTARD OR 'HO-MI Z' MUSTARD:** A slow-bolting mustard with crinkled, bright green leaves and deep purple veining, it has a sweet flavor with hint of spiciness.
- **'GIANT RED' MUSTARD:** A slow-bolting mustard with beautiful, large, crinkly red leaves; this variety is very spicy and crunchy.
- **'OSAKA PURPLE' MUSTARD:** This Japanese mustard has deep violet-purplish leaves, and green veins and stems.
- **'PURPLE WAVE' MUSTARD:** This open-pollinated mustard has light purple, frilly leaves, and is very hot and spicy.
- **'RED TATSOI' MUSTARD:** This variety grows to form beautiful red-purple rosettes, featuring crispy leaves with a mild mustard flavor.

RADICCHIO

Radicchio, also known as leaf chicory, is a prized garden vegetable renowned for its strikingly vibrant red leaves adorned with white veins. This versatile autumn and winter crop can be enjoyed raw or cooked, complementing other leafy greens. Radicchio has a unique flavor profile that pairs excellently with lettuce, making it a delightful addition to various culinary creations. Beyond its culinary value, radicchio is a nutritional powerhouse, rich in antioxidants, fiber, and vitamin K. These essential nutrients contribute to improved brain and heart health.

Planting

Radicchio is a cool-season crop that thrives in sunny to semi-shaded conditions. It can be direct-sown in the garden, seeded indoors, or transplanted as seedlings. Radicchio typically requires approximately 60 to 70 days to mature, significantly longer than lettuce and endive. Once established, radicchio can become a perennial.

Sowing Indoors

Sow radicchio seeds indoors approximately 8 to 10 weeks before the anticipated last frost date in the spring. Fill the seedling tray with a seed-starting mix and sow 2 to 4 seeds per cell, ensuring they are planted ¼ inch (0.6 cm) deep. Maintain even soil moisture during the germination period and position the seedling tray in a location where there are air temperatures from 55°F to 65°F (13°C–18°C). Alternatively, use a heat mat to help seed germination. Remove the heat mat once the seeds have germinated and relocate the tray under a grow light. Before transplanting, slowly acclimate the seedlings to outdoor conditions for at least 1 week.

Direct-Sowing

Radicchio is a hardy chicory that can tolerate cold temperatures. Sow radicchio seeds directly into the garden 2 to 4 weeks before the last frost in early spring, or sow the seeds in midsummer for a fall harvest. Plant radicchio 6 to 8 inches (15–20 cm) apart in the rows and 10 to 12 inches (25–30 cm) apart between the rows. Outside plant radicchio seeds ½ inch (1 cm) deep. The seeds should germinate within 6 to 10 days. Water the seeds well and maintain consistently moist soil. After the seeds have germinated, reduce the watering frequency.

Succession Planting

Succession planting is a gardening technique that involves sowing seeds every 1 to 2 weeks. This extends the harvest period.

Successful for most cool-weather crops, including radicchio, lettuce, carrots, spinach, radishes, peas, Swiss chard, and bok choy, succession planting helps avoid getting an overwhelming harvest with a large quantity of vegetables maturing all at once.

Care

Water

Radicchio, with its shallow roots, requires consistent watering of approximately 2 inches (5 cm) per week. Moisture is paramount for enhancing radicchio flavor. During hot summer days, radicchio thrives in afternoon shade. Excessive heat can cause the plant to bolt, resulting in bitter leaves. To mitigate bitterness, increase watering to 3 inches (7.5 cm) of water per week when nearing harvest time. Surround the base of the plant with organic mulch.

Blanching

Additionally, blanching the heart of the radicchio plant can enhance its sweetness. To blanch radicchio, simply shade the center of the mature plant from the sun by tying the leaves together. Do this on a dry day so the leaves don't rot. Another method is to cover the whole plant with a bucket. After 1 week, they can be harvested.

Fertilizer

Before planting, apply organic compost and a slow-release, all-purpose fertilizer. During the growing season, apply fish fertilizer every 3 weeks to support leaf growth.

Pests and Diseases

The most common pests and diseases for radicchio are aphids, slugs, snails, caterpillars, and powdery mildew. (See page 150.)

Companion Plants

Radicchio grows well with beets, carrots, asparagus, calendula, cilantro, eggplants, chives, garlic, mints, radishes, nasturtiums, onions, turnips, and parsnip. Avoid growing radicchio with brassicas and fennel.

Harvest and Storage

When to Harvest

If you are growing fall radicchio, allow the plants to remain until after the first few light frosts. The chilling temperatures will help reduce the bitterness.

Headed radicchio is ready when the entire head is firm and can be cut off at ground level. Loose-leaf radicchio can be harvested by cutting off the side leaves, leaving the center part to continue growing. Radicchio is a perennial, so after harvesting the head, the plant will regrow the following year.

How to Store

Radicchio, like other leafy greens, is not particularly shelf-stable and is best consumed fresh. Therefore, it is advisable to harvest only what you intend to use immediately. However, if you have an excessive harvest, it is possible to store radicchio for 7 to 10 days. To extend its shelf life, do not wash radicchio before storing it. Instead, place it in a plastic bag and store it in the refrigerator. When you are ready to use it, simply wash it with cold water and soak it for a few minutes.

My Favorite Vibrant Varieties

- **'GRUMOLO ROSSO' RADICCHIO:** A small, vigorous plant that forms a beautiful deep red rosette, it's a loose-leaf variety that has strong regrowth and can be grown as cut-and-come-again.
- **'PALLA ROSSA MELOT' RADICCHIO:** This radicchio has a round, dense head containing a flash of vibrant red leaves with white veins.
- **'ROSSA DI VERONA ARCA' RADICCHIO:** This solid-head radicchio has an elongated shape and deep red leaves with broad white veins.
- **'ROSSO DI CHIOGGIA' RADICCHIO:** This is the best-known radicchio with a round head that is dark red with white veins.
- **'TREVISO MESOLA' RADICCHIO:** From Italy, this radicchio has large heads of deep red leaves and white midribs.
- **'VARIEGATA DI CASTELFRANCO' RADICCHIO:** This radicchio forms a beautiful rosette of bright green leaves with red speckles. It's a cut-and-come-again type too.
- **'VARIEGATA DI CHIOGGIA' RADICCHIO:** This radicchio has a large, round head and leaves of red-pink and white variegation.

SPINACH

Spinach, a rapid-growing leafy green, is closely related to Swiss chard and beets. It is one of the most gratifying cool-weather vegetables to cultivate, yielding abundant harvests of tender leaves that can be enjoyed fresh in salads or incorporated into soups. Spinach is an exceptionally nutrient-dense food, containing a high concentration of antioxidants, iron, vitamins A, C, and K, folate, calcium, and fiber.

Planting

Spinach is a cool-season vegetable that thrives in sunny locations with a well-drained, fertile soil. It can be direct-sown in the garden or started indoors from seed and transplanted. Spinach has a growing time of approximately 1 month from seed to maturity. It can also be grown in containers.

There are two primary types of spinach: **savoy types**, characterized by large, puckered leaves, and **smooth-leaf types**, which have flat, upright leaves.

Direct-Sowing

In the spring, sow spinach seeds as soon as the soil becomes workable. For the fall, plant spinach seeds approximately 6 to 8 weeks before the anticipated first frost date.

Plant spinach seeds at a depth of ½ to 1 inch (1–2.5 cm) and at a spacing of 4 to 6 inches (10–15 cm) apart. Water thoroughly and maintain a consistent soil moisture. After the seeds have germinated, reduce the watering frequency. Spinach is an excellent crop for succession planting. Sow seeds every 1 to 2 weeks to ensure a continuous harvest.

Sowing Indoors

Approximately 6 weeks before the anticipated last frost date, sow spinach seeds indoors. Fill the seedling tray with a seed-starting mix and sow 2 to 4 seeds per cell, ensuring they are planted at a depth of ½ to 1 inch (1–2.5 cm). Maintain a consistent soil moisture level during the germination period. Additionally, place the seedling tray in a location with temperatures between 55°F and 65°F (13°C–18°C). If necessary, use a heat mat to help seed germination. Once the seeds have germinated, remove the heat mat and position the tray under a grow light. Spinach seeds typically germinate within 1 to 2 weeks.

Transplants

Spinach seedlings can be transplanted when they have reached a height of 3 to 4 inches (7.5–10 cm) or have developed at least two sets of true leaves. Before transplanting, acclimate the seedlings to outdoor conditions for 1 week. Plant the seedlings approximately 4 to 6 inches (10–15 cm) apart in rows, with an additional 8 to 10 inches (20–25 cm) of space between each row.

Care

Water

Spinach requires a substantial amount of water, approximately 1 to 1½ inches (2.5–4 cm) per week. Consistency is important too. To prevent leaf wetness, which can lead to plant diseases, it is advisable to water at the base of the plants. Additionally, mulch or a layer of garden straw can help in maintaining soil coolness and consistent moisture levels between watering sessions.

Fertilizer

Before planting, apply organic compost and a slow-release, all-purpose fertilizer to the spinach bed. Every 3 weeks during the growing season, apply a nitrogen-based fertilizer, such as fish fertilizer, to support foliage growth.

Pests and Diseases

The most common pests and diseases of spinach are aphids, slugs, snails, caterpillars, and powdery mildew. (See page 150.)

Companion Plants

Spinach grows well with beans, peas, radishes, kale, lettuce, chard, mustard greens, strawberries, and marigolds. Avoid growing spinach with potatoes, corn, and peppers.

Harvest and Storage

When to Harvest

Typically, spinach can be ready to harvest 4 to 6 weeks after planting seeds. However, you can harvest spinach as soon as the leaves are large enough to be consumed. Baby spinach leaves are sweet and tender. Pick individual leaves as soon as they reach the desired size, or you can cut the entire plant about 1 inch (2.5 cm) above the soil level, as spinach is a crop that can be harvested repeatedly.

How to Store

Spinach is at its most flavorful when consumed fresh. However, if there is an abundant harvest it can also be stored in a plastic bag in the refrigerator. Alternatively, blanch spinach leaves by cooking them in semi-boiling water for 30 seconds and then transfer them to an ice bath. Squeeze out the excess water and shape the spinach into balls about golf-ball size. Place these in ice cube trays or small muffin cups to freeze. Once frozen, store these spinach balls in a plastic bag in the freezer. Frozen spinach is best used within 3 to 6 months.

My Favorite Vibrant Varieties

- **'BLOOMSDALE LONG STANDING' SPINACH:** A fast-growing savoy type with large, round, glossy leaves that taste good even when the plant bolts.
- **'CATALINA' SPINACH:** A semi-savoy type with thick, succulent leaves.
- **'CELESTA' (F1) SPINACH:** A bolt-resistant variety that can be grown year-round.
- **'MEDANIA' SPINACH:** A slow-growing spinach with round and slightly curly leaves.
- **'SPACE' SPINACH:** A fast-growing hybrid with smooth leaves that is slow to bolt.
- **'TYEE' SPINACH:** A semi-savoy type with dark green leaves.

MALABAR SPINACH

If you are seeking a spinach alternative that will grow in the hot summer, consider cultivating malabar spinach. Malabar spinach is not a true spinach, but a succulent vine that thrives in hot summer weather. There are two primary varieties: red malabar spinach (*Basella rubra*) and green malabar spinach (*Basella alba*). I find that red malabar spinach is particularly appealing due to its vibrant red stems and heart-shaped, green foliage. The stems are adorned with purple flowers that eventually develop into purple berries. These edible berries are rich in betalain pigments, commonly used as a natural red dye.

Planting

Red malabar spinach is a resilient and heat-tolerant plant that can thrive in full sun, although it can also tolerate partial shade—which will result in larger leaves. It tolerates a range of soil conditions, but prefers well-draining, fertile, and high-organic matter soil. Red malabar spinach can be propagated from cuttings or seeds. It is a vining vegetable and can reach heights of up to 10 feet (305 cm), necessitating the use of trellising or a wire cage for support.

Direct-Sowing

In warmer climates, direct-sow malabar spinach seeds 3 to 4 weeks after the last frost date. If you garden in colder climates, commence seed germination indoors approximately 6 to 10 weeks before the last frost date. Transplant the seedlings to the garden once the soil has warmed and there is no likelihood of frost.

Sowing Indoors

For indoor sowing, fill the seedling tray with a seed-starting mix. Sow 1 to 3 seeds per cell, ensuring a depth of ½ to 1 inch (1–2.5 cm). Maintain even soil moisture and position the seedling tray in a location that is between 55°F and 65°F (13°C–18°C). Alternatively, you can use a heat mat during seed germination. Remove the heat mat after germination and place the tray under a grow light. Malabar spinach seeds typically germinate within 2 to 3 weeks, so exercise patience.

How to Propagate

Red malabar spinach is a relatively straightforward plant to propagate. To propagate the plant, simply sever a portion of the vine, trimming the bottom end at an angle, and immerse the angled end of the cutting in water.

Root formation typically occurs within 1 week. Once well rooted, transplant the new cutting into soil and maintain adequate moisture for a period of 2 weeks.

Care

Water

Red malabar spinach needs consistent watering to achieve optimal growth. Ensure that the soil remains consistently moist but is not waterlogged. Approximately 1 to 2 inches (2.5–5 cm) of water per week is required. Watering should be directed at the base of the plants. Organic mulch can be applied around the plants to retain soil moisture, suppress weeds, and regulate soil temperature.

Fertilizer

Fertilize malabar spinach with organic compost and slow-release, all-purpose fertilizer before planting and apply fish fertilizer every 3 to 4 weeks during the growing season to support vigorous growth.

Pruning

Pruning red malabar spinach is essential for maintaining its desired shape and height, and to promote dense growth. Regular pruning involves removing older leaves to encourage new growth and allow increased light and air to penetrate into the plant interior. Use clean garden shears or scissors to make precise cuts on individual leaves and stems. Avoid aggressive pruning, as this can result in a too-bushy plant. Red malabar spinach is a self-seeding plant that can easily spread too much or to places that you don't want it to grow. To prevent its spread, it is recommended to remove the berries before they mature.

Saving Seeds

Malabar spinach produces an abundance of edible, reddish-purple berries. These berries can be conveniently saved to start plants for next year or donated to a community garden or seed bank. Allow the berries to ripen until they attain a deep purple hue before harvesting them from the vine. Spread the berries on a paper towel and allow them to dry for several weeks or until they are completely dry. Once totally dry, store them in a paper bag.

Pests and Diseases

Red malabar spinach is generally resistant to pests and diseases but it can be susceptible to slugs, snails, and leaf spot. (See page 150.)

Companion Plants

Red malabar spinach grows well with peas, beans, chives, borage, basil, beets, Swiss chard, kale, lettuce, garlic, radishes, tomatoes, nasturtiums, and mustard greens. Avoid growing red malabar spinach with broccoli, cabbage, cauliflower, potatoes, Brussels sprouts, kohlrabi, and turnips.

Harvest and Storage

When to Harvest

Red malabar spinach reaches its full maturity within 65 to 80 days. To harvest, simply cut individual leaves, stems, and vine tips using a clean and sanitized pair of garden shears or scissors. Regular harvesting encourages the growth of more leaves. To harvest larger and more succulent leaves, wait until the plants reach their maturity.

How to Store

Red malabar spinach is most flavorful when consumed fresh. However, it can also be stored for later use. To store malabar spinach leaves, place them, without washing, in plastic bags and store them in the refrigerator for up to 1 week.

My Favorite Vibrant Varieties

- **GREEN MALABAR SPINACH (*BASELLA ALBA*):** Green malabar spinach has white or pink flowers, and bright green leaves with green vining stems.
- **RED MALABAR SPINACH (*BASELLA RUBA*):** Red malabar spinach has white or purple flowers, and bright green leaves with red vining stems.

↑ Red malabar spinach seedlings are ready to plant when they grow a couple sets of true leaves.

PURPLE BASIL

Purple basil is a visually appealing ornamental addition to any garden. It has deep purple leaves with jagged edges and a crinkled, textured appearance. It is good for culinary uses as the leaves impart a mild anise flavor reminiscent of sweet basil.

Planting

Purple basil, a semi-tropical herb, flourishes in warm, sunny environments. It does not tolerate any frost. Purple basil cultivation closely resembles that of green basil. It can be propagated from seeds and easily transplanted.

Sowing Indoors

In colder regions, indoor seed sowing extends the growing season. Sow purple basil seeds in multiple batches in seed-starting trays from late February through midsummer. After germination, relocate the seedlings to a warm and bright location, such as a windowsill. Purple basil seedlings can be transplanted once they have developed their first set of true leaves. Gradually acclimate the plants over a few weeks to outdoor conditions before transplanting.

Direct-Sowing

Purple basil can be sown outdoors once the temperature rises in early summer. Sow the seeds ⅛ inch (0.3 cm) deep, water gently, and maintain even soil moisture during the germination period. Purple basil seeds typically germinate within 1 week after sowing. Thinning is necessary for optimum healthy growth. Thinning involves plucking out the weaker, smaller seedlings

and leaving approximately 12 inches (30 cm) of space between each plant.

Care

Water

Regularly water purple basil with approximately 1 to 2 inches (2.5–5 cm) of water per week to ensure vigorous growth. The specific watering requirements are influenced by the soil type and temperatures.

Fertilizer

Purple basil is not a heavy feeder. Excessive fertilizer applications can negatively affect its flavor and aroma. Fertilizers with a high nitrogen content should be used to promote leaf growth. Apply a fertilizer once a month.

Pruning

Pruning purple basil plants stimulates both stem and leaf growth. The optimal time to commence pruning basil is when the plant has attained a height of at least 6 inches (15 cm) and has developed a few sets of true leaves.

Locate a suitable node—the junction where two sets of leaves emerge from the stem. Make a clean cut above the node, as this encourages the development of lateral shoots and results in bushier basil plants. However, avoid excessive pruning, as it can stress the plant and hinder its growth.

Pests and Diseases

The most common pests and diseases for purple basil are Japanese beetles, slugs, and root rot. (See page 150.)

Companion Plants

Basil grows well with tomatoes, peppers, oregano, marigolds, parsley, chives, borage, and lettuce. Avoid growing basil with sage, rosemary, thyme, fennel, and cucumber.

Harvest and Storage

When to Harvest

Purple basil can be harvested at any time by cutting the fresh, young leaves as needed.

If you prefer to harvest the entire stems, simply cut above a node. Regularly prune the younger leaves and stems before the flowers form to encourage new growth.

How to Store

Using basil when it is fresh and just harvested is the superior choice, as its tender leaves are a culinary delight. When in abundance, basil can be preserved through various methods, including drying, freezing, creating homemade pesto, or making basil salt.

My Favorite Vibrant Varieties

- **'AMETHYST IMPROVED' PURPLE BASIL:** A Genovese-type basil with foliage that is a stunning nearly black color.
- **'DARK OPAL' BASIL:** A fast-growing sweet basil with mostly purple leaves that are about 10% variegated green color; an open-pollinated, annual plant.
- **'RED RUBIN' BASIL:** An open-pollinated, annual plant with dark reddish-purple leaves.

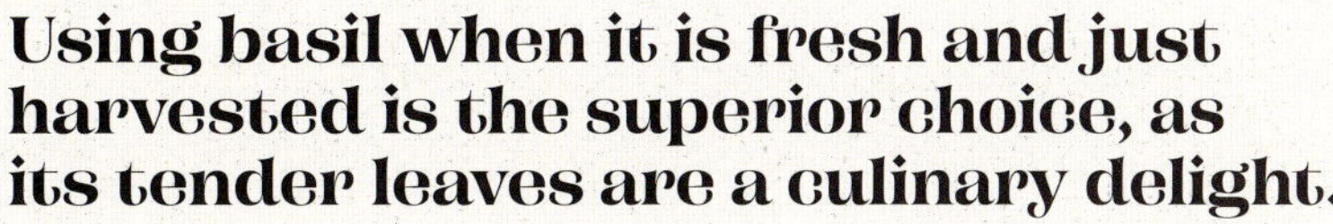
Using basil when it is fresh and just harvested is the superior choice, as its tender leaves are a culinary delight.

PURPLE SAGE

Purple sage is scientifically known as *Salvia officinalis*. 'Purpurascens' is a striking plant characterized by racemes of fragrant purple flowers and purple foliage with grayish-green undersides. Renowned for its dual ornamental and culinary value, purple sage holds a special place in the hearts of many cooks and gardeners alike due to its unique and delectable flavors.

Planting

Purple sage thrives in full sun and requires well-draining soil. It can be grown in containers or directly in the ground.

Transplants

The optimal time to plant purple sage is in the spring. I prefer growing purple sage from nursery transplants, as seeds can take a considerable amount of time to germinate. If planted in the ground, purple sage should be spaced at least 24 inches (61 cm) apart.

Care

Water

Purple sage requires minimal watering. Water the plant when the top 1 to 2 inches (2.5–5 cm) of soil has dried out.

Fertilizer

Purple sage is not a heavy feeder, and excessive fertilizer application can reduce its flavor. Therefore, it is recommended to apply compost or organic fertilizer only once every couple of months.

Pests and Diseases

Purple sage doesn't have any major pest or disease problems. Root rot can occur if the soil is too wet or powdery mildew from poor air circulation. (See page 152.)

Companion Plants

Sage grows well with rosemary, thyme, oregano, lavender, carrots, beans, and cabbage. Avoid growing sage with basil, cucumber, onions, and garlic.

Harvest and Storage

When to Harvest

The optimal time to harvest purple sage leaves is just before flowering. It is advisable to harvest the leaves individually or cut off the young shoots. However, it is crucial to avoid cutting off a lot of the woody plant parts.

How to Store

Fresh herbs are the optimal choice. However, to maximize your harvest, you can wrap fresh sage leaves with paper towels and store them in a plastic bag in the refrigerator for up to 2 weeks.

The summer months are an ideal time to dry fresh sage to ensure a ready supply during the rest of the year.

My Favorite Vibrant Varieties

- **PURPLE SAGE:** An attractive sage with purple foliage and lilac blue flowers, this variety grows to 12 to 18 inches (30–46 cm) tall and wide.
- **'TRICOLOR' SAGE:** A sage with striking multicolor grayish-green, white, and purple foliage, it provides great winter color in milder climates. It is more tender than other sage varieties and grows to 15 to 18 inches (38–46 cm) tall and wide.

VARIEGATED LEMON THYME

Variegated lemon thyme is a visually striking variety that enhances the aesthetics of any landscape with its vibrant yellow-cream variegated foliage and citrus aroma. Its compact growth habit, reaching a height of 6 to 10 inches (15–25 cm), makes it a low-growing, noninvasive, ideal groundcover for small spaces.

Planting

Variegated lemon thyme flourishes in well-lit, sunny locations. It thrives in slightly droughty, well-drained, alkaline to neutral soil and is not tolerant of wet conditions.

Variegated lemon thyme can be propagated through seeds or cuttings. Seed propagation is slow and irregular. Instead, purchase transplants or propagate from cuttings.

Sowing Indoors

In early spring, sprinkle the seeds evenly over the seed-starting soil surface and lightly dust the seeds with soil. Maintain even soil moisture, but avoid excessive sogginess.

Cuttings

Harvest suitable stems from established plants, cutting them approximately 3 to 4 inches (7.5–10 cm) in length. Remove the lower leaves. Fill small containers with potting mix and plant the cuttings about 1 to 2 inches (2.5–5 cm) into the soil. Water them deeply and keep them out of direct sunlight. Cover them with a humidity dome to maintain a moist environment. The cuttings should be rooted within 5 to 6 weeks.

Care

Water

Allow the top layer to dry out to a depth of 1 to 2 inches (2.5–5 cm) between waterings.

Fertilizer

Fish fertilizer can be applied once every 2 months.

Pruning

After the plant has completed blooming for the season, deadhead the flowers with hedge shears to encourage new growth.

Pests and Diseases

Variegated lemon thyme hardly has any problems with pests and diseases; however, there may be minor issues with aphids or spider mites. (See page 150.)

Companion Plants

Thyme grows well with rosemary, sage, lavender, strawberries, tomatoes, cabbage, eggplants, potatoes, shallots, blueberries, oregano, and parsley. Avoid growing thyme with mint, basil, cilantro, dill, chives, coriander, and fennel.

Harvest and Storage

Variegated lemon thyme can be harvested throughout the growing season. Harvest by snipping the new growth just above a node.

How to Store

Fresh herbs are the optimal choice. However, to maximize your harvest, you can wrap fresh thyme with a paper towel and store it in a plastic bag in the refrigerator.

Dry fresh thyme via hang drying, using a dehydrator, or simply sun drying.

My Favorite Vibrant Varieties

- **GOLDEN LEMON THYME:** The plant has golden and variegated leaves with a sweet lemon scent.
- **'MAGIC CARPET' CREEPING THYME:** This low-growing, aromatic groundcover matures in 75 to 90 days with purple-pink flowers.

Bud and Flower Vegetables

PURPLE SPROUTING BROCCOLI

Purple sprouting broccoli, a vibrant and nutritious vegetable belonging to the Brassicaceae family, exhibits a slightly sweeter flavor compared to its green counterpart. This unique variety produces numerous flower heads, albeit smaller in size, characterized by their tender texture. Purple sprouting broccoli is an exceptionally rich source of vitamins A, C, and K, as well as fiber, iron, and calcium. Furthermore, it contains antioxidants and phytochemicals that confer a range of health advantages.

Planting

Purple sprouting broccoli is a cool-season crop renowned for its exceptional hardiness and enhanced productivity. It is a winter and early spring crop. Cultivating purple sprouting broccoli is comparable to growing green broccoli. For optimal growth, purple sprouting broccoli thrives in a sunny location with partial shade and in fertile, well-drained soil, with a high organic matter content. Purple broccoli can be propagated from seed.

Sowing Indoors

To harvest purple sprouting broccoli in the spring, sow seeds indoors 6 to 8 weeks before the last frost date. Fill the seeding tray with seed-starting mix and sow 1 to 3 seeds per cell, ensuring they are planted ¼ to ½ inch (0.6–1 cm) deep. Maintain even soil moisture during the germination period and place the seeding tray in a location that has an air temperature between 60°F and 70°F (16°C–21°C). If necessary, use a heat mat to provide warmth during germination.

Purple broccoli seeds typically germinate within a 7-to-14-day period. Remove the heat mat once the seeds have germinated and position the tray under a grow light to prevent leggy seedlings.

Transplants

When the seedlings have grown sufficiently to handle, transplant them into larger pots. Broccoli seedlings are ready for transplanting when they have developed a couple of sets of true leaves, which usually takes approximately 5 to 8 weeks. Plant purple broccoli seedlings in raised beds or directly in the ground, spacing them 18 to 24 inches (46–61 cm) apart. If transplanting in containers, ensure that the container is at least 10 to 12 inches (25–30 cm) wide and deep.

Care

Water

Purple sprouting broccoli requires consistent watering, approximately 1 to 2 inches (2.5–5 cm) per week. Applying mulch around the plants also helps with water retention and suppressing weed growth.

Fertilizer

Purple sprouting broccoli is a heavy feeding crop. Apply nitrogen-based fertilizer during the initial transplanting stages of growth. Apply an additional feed of fish fertilizer and blood and bone meal once a month throughout the growing season.

Row Cover

Protect purple broccoli from pests such as aphids, cabbage worms, and flea beetles by covering the entire crop with lightweight floating row covers. Use PVC pipe hoops as the supporting structure.

Pests and Diseases

The most common pests and diseases of broccoli are aphids, cabbage worms, cabbage loopers, and leaf spots. (See page 150.)

Companion Plants

Purple sprouting broccoli and other brassicas grow well with onions, radishes, lettuce, beets, nasturtiums, chamomile, rosemary, sage, mint, and dill. Avoid growing purple broccoli with beans, eggplants, peppers, tomatoes, and potatoes.

Harvest and Storage

When to Harvest

Purple sprouting broccoli typically takes approximately 80 to 100 days to mature after planting the seed. Seed is usually planted in late spring or early summer for a winter harvest and late fall to early winter for a spring harvest. However, the maturity date can vary depending on the specific variety, and growing conditions.

Purple sprouting broccoli is ready for harvest when the small heads are well-formed but still tightly closed. The center head should be harvested first, followed by the side shoots, which can be harvested in the following weeks. The young leaves and stems are also edible. Purple sprouting broccoli exhibits a unique property; the more you harvest, the more the plants will produce.

Removing Bugs from Broccoli Heads

Fill a large container with warm water. Add some salt and baking soda. Ensure that the broccoli heads are completely submerged in the water. The water bath compels the insects to move out of the broccoli heads and washes out any insects that remain trapped.

How to Store

Purple sprouting broccoli is most suitable for immediate consumption after it is harvested. However, if an excess is available, it can be stored in a plastic bag in the refrigerator for a few days. Alternatively, blanch broccoli before storing it in the freezer.

My Favorite Vibrant Varieties

- **'BONARDA' PURPLE SPROUTING BROCCOLI:** A cold-hardy, F1 hybrid that matures in 210 days.
- **'BURGUNDY' SPROUTING BROCCOLI:** A cold-hardy broccoli that can grow year-round in mild climates.
- **'PURPLE MAGIC' BROCCOLI:** An F1 hybrid broccoli with a well-domed head and beautiful, deep purple, bright violet stems.

PURPLE BRUSSELS SPROUTS

Purple Brussels sprouts, a member of the Brassicaceae family, are renowned for their striking appearance and numerous health benefits. These diminutive purple cabbages are packed with essential vitamins, fiber, antioxidants, and other minerals, making them a nutrient-dense addition to any diet.

Planting

Purple Brussels sprouts are a cool-season vegetable and are slow-growing, so they require a long growing season. In my mild winter climate, I typically plant them in the fall and they grow through the winter months before they're harvested in the early spring. Brussels sprouts get better in flavor after being exposed to light frost.

Brussels sprouts thrive in full sun with at least 6 hours of direct sunlight—too much shade will slow their growth. They also require well-drained soil that is rich in organic matter and with a soil pH between 6.0 and 7.5. Purple Brussels sprouts can be grown from seed indoors or planted from nursery transplants.

Sowing Indoors

Sow seeds indoors 4 to 6 weeks before the recommended outdoor planting date. Fill the seedling tray with seed-starting mix and sow 1 to 3 seeds per cell at ¼ to ½ inch (0.6–1 cm) deep. Keep the soil evenly moist during the germination period and keep the seedling tray at an air temperature that is from 60°F to 70°F (16°C–21°C), otherwise you can use a heat mat during seed germination. Purple Brussels sprout seeds take about 7 to 14 days to germinate. Remove the heat mat after the seeds have germinated and place the tray under grow lights to prevent leggy seedlings. When the seedlings are large enough to handle, transfer them into a bigger pot. Purple Brussels sprout seedlings are ready to transplant when they grow a couple of sets of true leaves.

Transplants

Similar to other brassicas, the seedlings are ready to transplant when they have grown a couple sets of their true leaves—approximately after 5 to 8 weeks from sowing. Transplant purple Brussels sprout seedlings into a raised bed or in-ground garden. Space them 18 to 24 inches (46–61 cm) apart to give them room to grow and good airflow. Purple Brussels sprouts can be grown in containers too. Just make sure the containers are at least 12 inches (30 cm) wide and at least 12 inches (30 cm) deep.

↑ Purple Brussels sprout seedlings ready to transplant

↑ Purple Brussels sprouts 3 weeks after they have been transplanted

Care

Water

Purple Brussels sprouts need consistent watering with about 1 to 1½ inches (2.5–4 cm) of water per week. Apply mulch around the plants, which helps with water retention and reducing weed growth.

Packed with essential vitamins, fiber, antioxidants, and other minerals, making them a nutrient-dense addition to any diet.

Fertilizer

Purple Brussels sprouts are a heavy feeding crop. Apply nitrogen-based fertilizer during the initial stages of growth. Then apply an additional feed of blood and bone meal during the growing season.

Row Cover

To safeguard purple Brussels sprouts from pests such as flea beetles, cabbage worms, and aphids, it is advisable to cover the entire crop with a floating row cloth. Additionally, PVC pipe hoops can be used as a supporting structure for the row cloths. Row covers also serve as a barrier for bigger pests, preventing birds from damaging plants.

Pests and Diseases

The most common pests and diseases of purple Brussels sprouts are aphids, cabbage worms, cabbage looper, and leaf spots. (See page 150.)

Companion Plants

Purple Brussels sprouts and other Brassicaceae family members grow well with onions, radishes, lettuce, beets, nasturtiums, chamomile, rosemary, sage, mint, and dill. Avoid growing purple Brussels sprouts with beans, eggplants, peppers, tomatoes, and potatoes.

Harvest and Storage

When to Harvest

Each individual Brussels sprout grows out of the plant stalk right above a leaf joint. Smaller sprouts, when harvested, have more flavor. If the sprouts are left too long and get too large, they will taste like cabbage or will turn bitter. To boost the size of lower sprouts, remove the very top of the plant stalk to redirect plant energy to sprout development.

Use a sharp knife to remove the leaf below each sprout so the sprouts can be easily cut or pulled off the stalk. Another way to harvest purple Brussels sprouts, usually at the end of the season, is to cut the entire plant off at the soil surface.

How to Store

Purple Brussels sprouts are best processed within a few days after they are harvested. If delayed, they will lose sweetness, and their flavor will become strong. Do not wash the Brussels sprouts until you are ready to use them. Pat the sprouts dry, removing any dew or moisture, and store them in a plastic bag in the refrigerator.

You can also preserve them by using the blanching method. Partially cook Brussels sprouts in boiling water for about 4 minutes. Remove the sprouts from the pot and put them in ice water for another few minutes. Pat dry, spread out on a baking sheet, and freeze overnight. Once frozen, store the sprouts in a plastic bag or airtight container and keep them in the freezer for up to 6 months.

My Favorite Vibrant Varieties

- **'REDARLING' BRUSSELS SPROUTS:** This attractive Brussels sprout has purple-red heads, which still hold their color when you steam or roast them. It matures in 140 days.
- **'RED BULL' BRUSSELS SPROUTS:** This Brussels sprout with deep red-purple sprouts, leaves, and stems matures in 120 days.
- **'RED RUBINE' BRUSSELS SPROUTS:** Maturing in 90 days, this plant has deep purple sprouts and green leaves with purple veins and stems.
- **'ROSELLA' PURPLE BRUSSELS SPROUTS:** This purple Brussels sprout features oval-round leaves, with dark red sprouts and a milder, nutlike taste.
- **'TASTY NUGGETS' BRUSSELS SPROUTS:** These green sprouts with purple stems mature in about 78 days.

↑ Young sprouts starting to form at the leaf junctures after they have been transplanted

RAINBOW CAULIFLOWER

Rainbow cauliflower, a cool-weather crop belonging to the Brassicaceae family, is a beautifully vibrant type of cauliflower that enhances the vegetable garden much more than its white counterparts. It is available in various colors, including purple, orange, and green. Despite their distinct appearances, these varieties share similar cultural, taste, and nutritional profiles.

Among the purple cauliflower varieties, 'Purple of Sicily' and 'Graffiti' stand out as my personal favorites. Purple cauliflower exhibits a sweeter flavor and possesses a higher antioxidant content when compared to other varieties. The antioxidant anthocyanin, responsible for the purple pigment, contributes to the enhanced antioxidant properties.

Planting

Rainbow cauliflower thrives in sunny locations, with fertile, well-drained soil that is rich in organic matter. It can be direct-sown in the garden, or seeds can be started indoors, or seedlings purchased from the nursery and transplanted.

Sowing Indoors
To harvest rainbow cauliflower in the spring, sow seeds indoors 6 to 8 weeks before the last frost date. Fill the seeding tray with seed-starting mix and sow 1 to 3 seeds per cell, ensuring they are planted ¼ to ½ inch (0.6–1 cm) deep. Maintain even soil moisture during the germination period and place the seeding tray in a location with air temperatures between 60°F and 70°F (16°C–21°C). If necessary, use a heat mat to help seed germination. Similar to other brassicas, rainbow cauliflower seeds typically germinate within 7 to 14 days. Remove the heat mat once the seeds have germinated and position the tray under a grow light to prevent leggy seedlings. When the seedlings have grown sufficiently to handle, transplant

↑ Rainbow cauliflower, 'Graffiti' (purple), 'Romanesco' (green), and 'Cheddar' (orange) varieties

them into larger pots. Rainbow cauliflower seedlings are ready for transplanting when they have developed a couple of sets of true leaves.

Transplants
Seedlings are ready for transplanting when they have developed a few sets of true leaves, which usually takes approximately 5 to 8 weeks. Transplant cauliflower seedlings into raised beds or in-ground gardens and space them 18 to 24 inches (46–61 cm) apart. If transplanting into containers, ensure that the container is at least 10 to 12 inches (25–30 cm) wide and 12 inches (30 cm) deep.

↑ 'Romanesco' cauliflower

↑ 'Cheddar' cauliflower

Care

Water
Rainbow cauliflower needs consistent water of about 1 to 2 inches (2.5–5 cm) per week. Applying a mulch around the plants also helps with water retention and will reduce weed growth.

Fertilizer
Rainbow cauliflower is a heavy feeding crop. Apply a nitrogen-based fertilizer during the initial stages of growth. Later apply additional feedings of blood and bone meal regularly throughout the growing season.

Row Covers
To safeguard rainbow cauliflower from pests such as flea beetles, cabbage worms, cabbage loopers, and aphids, it is advisable to cover the entire crop with a floating row cloth. PVC pipe hoops can be used as a supporting structure for the row cloth. Row covers will also serve as a barrier for bigger pests, preventing birds from damaging plants.

Blanching
Blanching is a technique employed in the garden to shield cauliflower heads from direct sunlight exposure. This is particularly beneficial for white cauliflower, as sunlight can cause the head to turn yellowish cream in color and compromise its flavor. The blanching process involves using the large outer leaves of the cauliflower plant. Once the cauliflower head has reached a diameter of a few inches, the leaves are tied or clipped together to create a barrier that prevents sunlight from reaching the head. Additionally, this method serves as a deterrent against birds pecking at the cauliflower.

Many newer cauliflower varieties possess a self-blanching trait, meaning that the upper leaves naturally fold over the center of the plant, effectively blocking sunlight.

Pests and Diseases

The most common pests and diseases for cauliflowers are flea beetles, aphids, cabbage worms, cabbage looper, and leaf spot. (See page 150.)

Companion Plants

Rainbow cauliflower and other brassicas grow well with onions, radishes, lettuce, beets, nasturtiums, chamomile, rosemary, sage, mint, and dill. Avoid growing cauliflowers with beans, eggplants, peppers, tomatoes, and potatoes.

Harvest and Storage

When to Harvest

Harvest rainbow cauliflower heads as soon as they reach a diameter of 6 to 8 inches (15–20 cm) and are still firm and compact. Use a sharp knife to sever the cauliflower head from the stalk, leaving a few leaves attached. This method will help preserve the cauliflower's freshness in the refrigerator for an extended period.

Removing Bugs from Cauliflower Heads

Fill a big bucket with water and add some salt and baking soda. Make sure to fully submerge the cauliflower head, as the water bath helps by both forcing the pests out of the head and killing any insects that are trapped inside.

How to Store

Rainbow cauliflowers are best eaten fresh, but they can be stored in the fridge for up to 10 days. If you have an abundant harvest, it is important to store your cauliflower properly. One of the best ways to store and preserve cauliflower is to freeze it. Blanching the individual cauliflower florets will extend their freezer shelf life. To blanch, boil the cut cauliflower florets in salted water for a few minutes. Then, drain and rinse them in cold water. Store the florets in a sealed, clear, plastic bag in the freezer.

↑ Cauliflower plants before they develop their distinctive heads

My Favorite Vibrant Varieties

- **'CHEDDAR' CAULIFLOWER:** An earlier cauliflower, maturing in 60 to 70 days with a bright orange head, the heads are firm with dense, smooth curds, and a diameter of about 5 to 7 inches (13–18 cm).
- **'GRAFFITI' PURPLE CAULIFLOWER:** This cauliflower has a large and stunning dark purple head that matures in 80 days.
- **'PURPLE OF SICILY' CAULIFLOWER:** This cauliflower has a tightly compact, brilliant purple head. It will mature to have a 2 to 3 pound (1 kg) head with sweet flavor in 70 days.
- **'ROMANESCO' CAULIFLOWER:** This plant is a cross between a cauliflower and broccoli, with lime-green curds that are unusually pointy and whorls that develop in a fractal pattern. It takes about 85 days to mature, with heads that measure approximately 6 to 7 inches (15–18 cm) in diameter.

CALENDULA

Calendula, commonly known as pot marigold, is an annual herb belonging to the Asteraceae family. It produces vibrant orange, yellow, white, and bicolored flowers, depending on the variety. Planting calendula attracts pollinators and beneficial insects to the vegetable garden and is one of the best companion plants to deter unwanted pests.

Planting

Calendula prefers a sunny to semi-shady location and nutrient-rich soil. It can be easily propagated from seeds indoors or direct-sown outdoors in the garden. It also thrives in containers.

Direct-Sowing
Sow calendula seeds after the final frost of the spring season or as soon as the soil becomes workable. Direct-sow the seeds in the garden, in raised beds, or in containers at a depth of ½ inch (1 cm) and at a spacing of 6 to 12 inches (15–30 cm) apart, depending on the variety. Maintain even soil moisture until the seeds germinate, usually within 6 to 10 days.

Sowing Indoors
Start calendula seeds indoors approximately 8 weeks before planting them outdoors in the garden. Fill the seeding tray with a seed-starting mix and sow 1 to 2 seeds per cell, ensuring a depth of approximately ½ inch (1 cm). Maintain an adequate soil moisture during the germination phase. Additionally, position the seeding tray in a location that has an air temperature from 60°F to 70°F (16°C–21°C). If necessary, use a

heat mat to help seed germination. Once the seeds have germinated, remove the heat mat and provide them with indirect sunlight to prevent leggy seedling growth. Calendula seedlings are ready for transplanting when they have developed a couple of sets of their true leaves. Before planting them in the garden outdoors, acclimate the seedlings to outdoor conditions over a period of approximately 1 week, a little more each day.

Care

Water

During the peak of the summer season, water calendula once or twice a week. It is crucial to avoid watering the plant excessively, as calendula dislike having their leaves wet. Excessive moisture can lead to fungal diseases like powdery mildew.

Fertilizer

Calendula are not heavy feeders, so fertilize them with a well-balanced fertilizer once a month.

Pests and Diseases

The most common pests and diseases for calendula are aphids, caterpillars, spider mites, and powdery mildew. (See page 150.)

Companion Plants

Calendula grows well with tomatoes, peppers, carrots, asparagus, cucumbers, strawberries, beans, lettuce, spinach, broccoli, cauliflower, cabbage, and kale. Avoid growing calendula with potatoes, garlic, onions, and fennel.

↑ Harvested calendula flowers

Harvest and Storage

When to Harvest

Calendula plants will produce a significant number of blooms throughout the growing season. To help them along and to ensure a continuous supply of flowers throughout the summer, provide them with adequate water and nutrients, and regularly harvest the flowers.

Regularly deadheading the spent flowers will encourage the plants to produce new blooms. If the plant begins to wilt, cut it back severely. This will help the plant recover and put out new healthy growth.

Planting calendula attracts pollinators and beneficial insects to the vegetable garden and is one of the best companion plants to deter unwanted pests.

CALENDULA BENEFITS AND USES

Calendula has a longstanding reputation as a traditional remedy that is naturally anti-inflammatory and antifungal. It is used for treating skin conditions, such as eczema, and healing wounds.

Calendula flowers are edible and commonly harvested to use in tea, oil, and vinegar infusions. Calendula tea is known for its soothing properties and can aid digestion.

Saving Seeds

Allow some of the flowers to go to seed so you can save the seed for next year. Once the flower heads turn brown, they will develop thick, crescent-shaped seeds. Cut the seedheads off and dry them in a well-ventilated area for a few weeks until they are fully dry. Gently break apart the flower head and store the dry seeds in a glass jar or paper envelope. Then put the seeds in a cool, dry, and dark place for long-term storage.

My Favorite Vibrant Varieties

- **'BALL'S IMPROVED ORANGE' CALENDULA:** Flowering just 45 to 55 days from seeding, this calendula is a beautiful deep orange color.
- **'ORANGE KING' CALENDULA:** This calendula produces massive, double, brilliant orange flowers starting 65 days from seeding.
- **PLAYTIME MIX CALENDULA:** Producing an amazing splashy mix of colors from warm tones and creamy pastels to bright orange, this calendula blooms 45 to 65 days from seeding.
- **'PINK SURPRISE' CALENDULA:** After only 45 to 55 days from seeding, this calendula has apricot flowers tinged with pink.

CALENDULA-INFUSED OIL

↑ Using dried calendulas to make soaps and balms

Calendula-infused oil has many beneficial properties for use as a natural skin care product. Many home skin care recipes start with a good calendula-infused oil, including those for making natural soaps and salves. Infused oil will last up to the expiration date on the original oil package label. Good oils to use include coconut oil, sweet almond oil, grapeseed oil, jojoba oil, and olive oil. Always use completely dried calendula petals.

Choose one of the below methods to make the infused oil, then remove the cheesecloth containing the herbs and store the infused oil in a cool and dark place. There are three methods to make calendula oil.

Sun

Pack the dried petals in Mason jars and pour your chosen type of oil over them. Close the lid and leave it in the sun for 8 to 10 hours, avoiding intense sunlight because overheating the oil can deplete some of the beneficial properties of the calendula. Alternatively, if using an oil that remains liquid at room temperature, you can place the jar on a warm, sunny windowsill and shake it at least once a day. Leave it on the windowsill to infuse for approximately 4 to 6 weeks.

Stovetop

This method uses a double boiler to gradually heat the oil and dried calendula blooms. Ensure that the heat is low and infuse for 3 to 5 hours. You will notice that the oil will transform into a visually appealing golden color and become fragrant.

Slow Cooker

The petals will infuse into the oil at the lowest heat setting—which can take from 8 to 12 hours. Place the calendula flowers in the slow cooker and completely cover with the oil of your choice. Be sure to leave the lid off, as condensation on the lid can drip into the pot, potentially introducing mold and bacteria.

MARIGOLDS

Marigolds are among my preferred annual flowers to cultivate during the summer months. Their diverse range of colors renders them an ideal addition to any garden or bouquet. Marigolds possess remarkable abilities to attract pollinators and beneficial insects, making them an exceptional companion plant for garden cultivation. These vibrant flowers also hold significant cultural importance around the world.

Planting

Marigolds are annuals that typically bloom continuously throughout the summer months until frost arrives. They can be direct-sown outdoors or started indoors, and they germinate relatively quickly. Marigolds thrive in full sun.

There are three types:

- **African marigolds** produce large blooms and can reach heights of up to 4 feet (122 cm).
- **French marigolds** grow between 8 and 12 inches (20–30 cm) tall.
- **Signet marigolds** are less than 6 inches (15 cm) tall with small flowers.

Direct-Sowing

Sow marigold seeds directly into the ground at a depth of ¼ inch (0.6 cm) and a spacing of 6 to 12 inches (15–30 cm) apart. Maintain even soil moisture until the seeds germinate, in 6 to 10 days.

Sowing Indoors

Sow marigold seeds indoors 5 to 8 weeks before the last frost date. Fill the seedling tray with a seed-starting mix and sow 1 to 2 seeds per cell, at a depth of ¼ to ½ inch (0.6–1 cm). Maintain adequate moisture and position the seedling tray in a location with air temperatures between 60°F and 70°F (16°C–21°C). If necessary, use a heat mat to help seed germination. Remove the heat mat once the seeds have germinated and provide them with indirect sunlight from a grow light. Transplant outdoors when they have developed a couple of sets of true leaves.

Care

Water

Marigolds can tolerate a bit of drought. Water marigolds deeply once a week and allow the soil to dry out between waterings.

Fertilizer

Fertilize marigolds with a well-balanced organic fertilizer once every 3 to 4 weeks.

Pests and Diseases

The most common pests and diseases for marigolds are aphids, spider mites, slugs, snails, and powdery mildew. (See page 150.)

Companion Plants

Marigolds grow well with tomatoes, peppers, potatoes, eggplants, cucumbers, zucchini, melons, squash, beans, lettuce, kale, broccoli, cauliflower, cabbage, and strawberries. Avoid growing marigolds with pole beans and cabbage.

Harvest and Storage

When to Harvest

Regularly deadhead the old marigold flowers to encourage the plants to produce new blooms by cutting the stem below the finished flower, just above a node (similar to pruning basil). New shoots will emerge from the nodes.

My Favorite Vibrant Varieties

- **'KEES' ORANGE' MARIGOLD:** A massive and magnificent African marigold with bright orange blooms.
- **'QUEEN SOPHIA' MARIGOLD:** Large, showy double blossoms with russet red and gold petals.
- **'STRAWBERRY BLONDE' MARIGOLD:** A French marigold with bicolor pastel pink, rose, and yellow blooms.

ZINNIAS

Zinnias, a vibrant and prolific annual, thrive throughout the summer months, captivating gardeners with their array of colors. In the Asteraceae family, these plants are originally native to Mexico and Central America. The flowers offer a diverse range of hues, making them an ideal addition to any garden or indoor bouquet.

For novice gardeners, zinnias present an exceptional opportunity due to their ease of cultivation and maintenance. Their blooms also attract pollinators such as hummingbirds and butterflies, contributing to the overall biodiversity of the garden.

Planting

Zinnias are annuals that are available as dwarf varieties, which are low-growing and compact, while some can reach considerable heights of up to 4 feet (122 cm). Zinnias can be direct-sown outdoors in the garden or started from seed indoors. For the taller varieties, trellising or staking may be necessary to prevent them from toppling over in heavy rain or wind.

Direct-Sowing

Zinnias flourish in sunny locations with nutrient-rich, well-draining soil. Plant zinnia seeds directly in the ground, in raised beds, or in containers at a depth of ¼ inch (0.6 cm) and at a spacing of 6 to 12 inches (15–30 cm), depending on the variety. Maintain a consistent soil moisture until the seeds germinate; usually this is within 6 to 10 days.

Sowing Indoors

Start zinnias seeds indoors 5 to 8 weeks before the last frost date. Fill the seedling tray with seed-starting mix and sow 1 to 3 seeds per cell and ¼ to ½ inch (0.6–1 cm) deep. Keep the soil moist during the germination period and keep the seedling tray in a location with temperatures from 60°F to 70°F (16°C–21°C), otherwise you will need a heat mat for good seed germination. Remove the heat mat after the seeds have germinated and place the seed tray under grow lights to prevent leggy seedlings. The zinnia seedlings are ready to transplant when they have grown a couple of sets of true leaves. Harden them off for about 1 week so they are ready for outdoor conditions.

↑ Beautiful colorful zinnias in south-side garden

Care

Water

Keep zinnias consistently watered during hot, dry weather. Zinnias do not like their leaves wet, so try not to get water on the foliage or use overhead irrigation. Prolonged wet foliage can cause fungal diseases such as powdery mildew.

Fertilizer

Zinnias are not heavy feeders, so fertilize them with a well-balanced fertilizer once a month.

Pests and Diseases

The most common pests and diseases for zinnias are aphids, caterpillars, spider mites, and powdery mildew. (See page 150.)

Companion Plants

Zinnias grow well with tomatoes, peppers, cucumbers, eggplants, marigolds, basil, sunflowers, beans, lettuce, corn, carrots, and nasturtiums. Avoid growing zinnias with potatoes, cucumbers, dill, and brassicas like cabbage, broccoli, and cauliflower.

↑ Zinnias make a colorful bouquet.

Harvest and Storage

When to Harvest

Zinnias have a long blooming period. To ensure continuous flower production, provide them with adequate water throughout their growing season. Harvest zinnia flowers when the petals are fully open, and when the stem has become rigid. Zinnias can last up to 12 days in a vase, provided the water is changed every few days.

Saving Seeds

Regularly deadhead the spent flowers to encourage the plants to produce new blooms. Cut the stem just above a node (similar to pruning basil) and new branches will emerge at the node.

Once the flower heads turn brown, cut them off and allow them to dry in a well-ventilated area for several weeks or until they are completely dry. Gently break apart the seedhead to reveal the arrowhead-shaped structures attached to the shriveled petals. Allow these structures to dry for a few more days to ensure they are completely dry. Store the dried seeds in a glass jar or paper envelope and place them in a cool, dry, and dark location.

My Favorite Vibrant Varieties

- **'BENARY'S GIANT CORAL' ZINNIA:** Blooming after only 60 to 75 days from seed, this zinnia has large, radiant coral-pink blooms.
- **'MAZURKIA' ZINNIA:** This dahlia-type zinnia has large blooms after 60 to 75 days from seeding. Each bloom has scarlet petals tipped in cream.
- **'PEPPERMINT STICK MIX' ZINNIA:** Colorful, medium-sized blooms come in a mix of colors with striped and speckled petals. Blooming starts 75 to 90 days from seeding.
- **'QUEEN LIME BLUSH' ZINNIA:** This zinnia produces large blooms 75 to 90 days from seeding that are vibrant lime green to red ombre with red centers.
- **'REDMAN SUPER CACTUS' ZINNIA:** This huge, red-flowered, giant cactus-type blooms from 70 days after seeding.
- **'SEÑORITA' ZINNIA:** This variety features enormous cactus-type blooms from salmon to cheery peachy pink colors. Blooms 60 to 75 days from seeding.

For novice gardeners, zinnias present an exceptional opportunity due to their ease of cultivation and maintenance.

Fruiting Vegetables

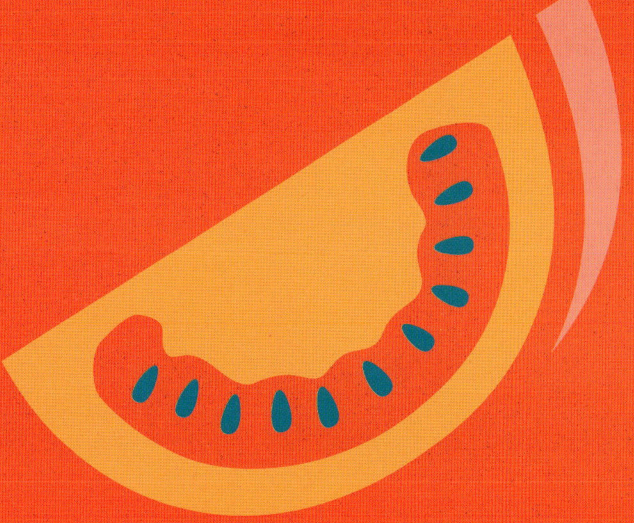

YELLOW, ORANGE, AND WHITE CUCUMBERS

Cucumbers are one of the oldest fruits cultivated globally. Having adapted to various climates around the world, many unique varieties of cucumbers have been developed. They exhibit a diverse range of shapes, sizes, colors, and flavors. Cucumbers belong to the Cucurbit family and are relatively easy to cultivate. They exhibit rapid growth and produce a substantial yield of fruits. Cucumbers contain many important vitamins and phytonutrients, and eating them is a great way to stay hydrated.

Planting

Cucumbers are heat-tolerant plants that flourish in late spring and summer, but they are not frost resistant and don't like cold. Plant cucumber seeds when the soil temperature reaches at least 70°F (21°C). If you are starting seeds indoors, sow them approximately 3 weeks before outdoor planting. However, cucumbers prefer direct-sowing into the outdoor garden soil, as they dislike being transplanted and having their roots disturbed. Cucumbers are available in two varieties: **bush cucumbers** and **climbing cucumbers.**

Direct-Sowing

Cucumbers flourish in full sun. Direct-sow cucumber seeds after the last frost date in the spring and when the soil has warmed sufficiently. Sow the seeds 1 inch (2.5 cm) deep and approximately 2 to 3 feet (61–91 cm) apart. Water the seeds thoroughly and maintain consistently moist soil.

Care

Watering

Cucumbers require approximately 1 to 2 inches (2.5–5 cm) of water per week. Water deprivation can result in a bitter cucumber flavor at harvest time. To enhance soil moisture retention, apply organic mulch around the base of the plant.

Fertilizer

Fertilize the plants with a well-balanced fertilizer once they have commenced flowering. Follow the instructions provided on the fertilizer label.

Pruning

Pruning vining cucumber types will enhance their yield and improve airflow, which mitigates the risk of fungal diseases. Furthermore, good airflow around the plants helps insects find and pollinate the blossoms.

Pollination

Certain cucumber varieties are self-pollinating, while others will need manual pollination to produce fruit. In the absence of natural pollinators like bees, hand pollination can be done to aid in this process. Pollination failure may result in deformed cucumbers, slow growth, or even the complete absence of fruit. While hand pollination can be time-consuming, it is an effective method for achieving a substantial crop yield.

The most important aspect of hand pollination is to distinguish between the male and female flowers, which both develop on the same plant. Male flowers have a short stem, while female flowers

bear a small fruit at their base. In the early morning when flowers have opened, use a small paintbrush or cotton swab to collect pollen from the male flowers. Then, gently transfer this pollen to the center of the female flowers.

Pests and Diseases

Cucumbers are susceptible to various pests and diseases, including beetles, leaf miners, and powdery mildew. (See page 150.)

Companion Plants

Cucumbers grow well with dill, nasturtium, marigolds, corn, beans, radishes, borage, lettuce, onions, oregano, sunflowers, lavender, peas, and chamomile. Avoid growing cucumbers with potatoes, mint, fennel, sage, and squash.

Harvest and Storage

When to Harvest

Cucumbers typically require approximately 50 to 70 days from seed, and this duration can vary depending on the specific variety and weather conditions. To ensure that cucumbers are harvested at the optimal time, be aware of the mature size of the cucumber varieties you are cultivating.

Cucumber fruits grow rapidly, so regularly inspect the plant for ripe fruits. If the fruits are left on the vine for too long, the results will be oversized fruits that may become bitter, and this will signal the plant to stop future production. Use sharp scissors or pruners to snip the stems rather than attempting to pull the cucumbers off the plants.

GROWING CUCAMELON

Cucamelons are relatively easy to cultivate and exhibit remarkable productivity, yielding hundreds of diminutive fruits.

Start sowing cucamelon seeds indoors approximately 6 to 8 weeks before the anticipated last frost date in the spring. Sow 1 seed per cell and ensure a depth of 1 inch (2.5 cm). Maintain even soil moisture during the germination phase. Additionally, place the seedling tray where air temperatures are between 60°F and 70°F (16°C–21°C). If necessary, use a heat mat to help seed germination.

- **Sunlight:** Cucamelons require full sun and protection from strong winds.
- **Soil:** Cucamelons prefer a well-drained and fertile soil.
- **Watering:** Cucamelons are drought-tolerant, but the soil should be consistently moist. Water at the base of the plant to avoid overhead watering, which can lead to fungal diseases.
- **Fertilizer:** Regularly feed cucamelons with a high-potassium fertilizer every week during the flowering stage.
- **Harvest:** Cucamelons are ready for harvesting when the fruits are approximately the size of a large grape and are still firm.

How to Store

Cucumbers can be stored at room temperature for up to 1 week. For longer storage, refrigerate cucumbers. Another method of preserving cucumbers is to pickle them. Dill pickles are a popular choice and a delicious way to preserve cucumbers.

My Favorite Vibrant Varieties

- **'DRAGON'S EGG' CUCUMBER:** This cucumber has a beautiful cream color and is shaped like an egg. It is sweet tasting, with a maturity of 55 to 65 days.
- **'LEMON' CUCUMBER:** This cucumber has the shape and color of a lemon, but with a super sweet taste. It matures in 60 days.
- **'SIKKIM' CUCUMBER:** This cucumber with large fruit has a unique rusty red color and matures in 65 to 75 days.

MULTIHUED EGGPLANTS

Multihued eggplants are a captivating ornamental addition for any summer garden. One of the most attractive is the 'Fairy Tale' eggplant. These miniature fruits, adorned with lavender hues and striking white streaks, create a visually enchanting display. 'Fairy Tale' eggplant is a relatively fast-growing plant, reaching maturity within 50 days and producing clusters of fruits. They have a sweet taste and a firm texture.

Planting

Multihued eggplants are a great choice for the summer vegetable garden, and often they are also well-suited (like the 'Fairy Tale' eggplants) for containers or raised beds. Multihued eggplants are easy to grow and can be planted from seeds or transplanted into the garden. They prefer a location with full sun.

Sowing Indoors
Fill the seedling tray with a seed-starting mix and sow 1 to 2 seeds per cell, ensuring a depth of approximately ½ inch (1 cm). Maintain adequate soil moisture during the germination phase. Place the seedling tray in a location where air temperatures are between 75°F and 85°F (24°C–29°C). If necessary, use a heat mat to help seed germination. Once the seeds have germinated, remove the heat mat and provide them with indirect sunlight to prevent leggy seedlings. The seedlings are ready for transplanting when they have developed a couple of sets of true leaves.

Transplants
Multihued eggplants are highly susceptible to cool temperatures, so transplant seedlings outdoors when the air tem-

These miniature fruits, adorned with lavender hues and striking white streaks, create a visually enchanting display.

perature exceeds 60°F (16°C). Before moving them outdoors, acclimate the seedlings to outdoor conditions slowly over a period of approximately 1 week. Space the plants approximately 24 inches (61 cm) apart or plant them in a container.

Care

Water

Consistent watering is crucial for the development of fruit. Eggplant fruit will be bitter if the plant does not receive adequate water. Give eggplants at least 1 inch (2.5 cm) of water per week. During the hottest months, increase watering to 2 to 3 times per week.

Fertilizer

Multihued eggplants are heavy feeders and require regular fertilization. Apply a well-balanced fertilizer every 3 to 4 weeks, and additionally side-dress with a high phosphorus fertilizer to encourage fruiting.

↑ A pretty little eggplant harvest

Pruning and Support

'Fairy Tale' eggplant is a dwarf variety, but other eggplant varieties can grow quite tall. To prevent the tall varieties from toppling over, ensure you provide enough support like stakes or tomato cages. Regularly prune any weak growth to promote healthy, bushy eggplants and to encourage fruiting.

Companion Plants

Multihued eggplants grow well with marigolds, pole beans, borage, nasturtiums, spinach, oregano, lettuce, peas, basil, mint, broccoli, chives, dill, fennel, onions, thyme, beets, calendula, and catnip. Avoid growing eggplants with potatoes (though sweet potatoes are acceptable), peppers, tomatoes, and tomatillos.

Pests and Diseases

Multihued eggplants are susceptible to flea beetles, cutworms, and caterpillars. Additionally, they are prone to diseases such as powdery mildew. (See page 150.)

Harvest and Storage

When to Harvest

Multihued eggplants typically mature during mid-to-late summer, depending on the specific variety. If you are propagating them from seeds, the cultivation process will take approximately 100 to 120 days, whereas transplants will require around 60 to 80 days to mature.

Eggplants are ready to be harvested when their skin is glossy and devoid of wrinkles. For optimal flavor, it is best to pick the fruit when it is young. Use a sharp knife or clean pruners to sever the fruit from the plant when harvesting.

How to Store

Eggplant is at its peak flavor when freshly harvested and can be enjoyed raw, baked, grilled, or stir-fried. If you have an excess harvest, they can be stored in the refrigerator for 1 week.

My Favorite Vibrant Varieties

- **'CASPER' EGGPLANT:** A medium-sized eggplant with a smooth ivory skin, it matures in 75 days from seeding.
- **'FAIRY TALE' EGGPLANT:** This miniature eggplant with clusters of elongated lavender and white-striped fruit matures from seed in 65 days.
- **'LISTADA DE GANDIA' EGGPLANT:** This eggplant with bright purple white stripes has a fabulous flavor with sweet, tender flesh. It's ready to harvest 90 days from seeding.
- **'MELANZANA ROSSA DI ROTONDA' EGGPLANT:** This unique variety tastes like a combination of eggplant, tomato, and pepper. It has red-orange, round fruits with green stripes and matures in 70 to 80 days.
- **'ROSA BIANCA' EGGPLANT:** This sweet and tender eggplant has light pink-lavender fruit and white shading. Harvest it 80 days from seeding.

MELONS

Melons provide a delightful and visually appealing addition to summer gardens. They belong to the Cucurbitaceae family, which also includes squash and cucumbers. Cultivating melons is an enjoyable and rewarding experience, as they yield an abundance of flavorful, juicy, and sweet summer treats.

Planting

Melons thrive in full sun and well-drained soil. Therefore, it is advisable to plant them in raised beds in a location with ample space for the vines to spread.

Direct-Sowing

Plant melon seeds directly in the garden approximately 10 to 14 days before your average last frost date or wait until the soil reaches a suitable temperature over 70°F (21°C). To determine soil temperature, use a thermometer to measure the temperature within the top 2 to 3 inches (5–7.5 cm) of soil. Melon seeds germinate best at soil temperatures from 75°F to 85°F (24°C–29°C).

At planting time, generously apply organic matter and an all-purpose fertilizer. Sow 1 to 3 seeds at a depth of 1 inch (2.5 cm) and maintain a spacing of 18 to 24 inches (46–61 cm) between each plant in the row. Space rows 5 to 6 feet (152–183 cm) apart. Ensure adequate soil moisture until the seeds germinate.

Sowing Indoors

Melons are highly susceptible to issues with root disturbance when transplanted. This, and their preference for warm temperatures, often necessitates indoor seed sowing. Sow seeds indoors approximately 2 to 4 weeks before the anticipated last spring frost date. Fill the seedling tray with a seed-starting mix and sow 1 seed per cell, ensuring a depth of approximately 1 inch (2.5 cm). Maintain adequate and consistent soil moisture. Place the seedling tray where the room air temperature is from 65°F to 75°F (18°C–24°C). If necessary, use a heat mat to help seed germination. Once the seeds have germinated, remove the heat mat and provide them with indirect sunlight to prevent leggy seedlings. Seedlings are ready for transplanting once they have developed a few sets of their true leaves.

Care

Water

Melons require consistent moisture, approximately 1 to 2 inches (2.5–5 cm) of water per week. Excessive watering can cause fruit splitting, so gradually reduce watering as the fruit ripens. This will also improve the flavor. To enhance soil moisture retention, suppress weeds, and regulate soil temperature, apply organic mulch around the base of the plant.

Fertilizer

Melons prefer a high-nitrogen fertilizer to encourage vine growth at the beginning of the season. Once the vines begin to produce flowers, fertilize them with a fertilizer that is high in phosphorus and potassium (but low in nitrogen), to help fruit development.

Hand Pollinating

Melons produce both male and female flowers on the same plant. To maximize fruit production, it is beneficial to hand pollinate the flowers. The most important aspect of hand pollination is to distinguish between the male and female flowers.

Male flowers have a short stem, while female flowers bear a small fruit at their base.

In early morning when the flowers have just opened, select a healthy male flower and carefully remove the petals to expose the interior of the bloom. Gently transfer this pollen to the center of the female flowers.

Thinning Fruits

To cultivate the most delectable and nutritious melons, it is imperative to thin out the developing fruits and allow only a limited number to grow on each plant. It is often necessary to put up a netting system to provide support for the heavy fruits. Some gardeners use pantyhose to support individual melons on a trellis.

Companion Plants

Melons grow well with marigolds, radishes, nasturtiums, beans, oregano, lavender, borage, and sunflowers. Avoid growing melons with cucumbers, zucchini, pumpkins, and winter squash.

Pests and Diseases

The most common pests and diseases of melons are aphids, beetles, cutworms, white flies, and powdery mildew. (See page 150.)

Harvest and Storage

When to Harvest

Harvesting time for melons can differ depending on the specific variety. A ripe melon will release a sweet aroma and be a bit soft at the stem end. The skin may also change color, and you will observe some minor cracks forming around the stem.

How to Store

Different types of melons require specific storage temperatures to maintain their quality. While most melons can be stored at room temperature for a few days, it is recommended to refrigerate them until they are ready to be consumed.

My Favorite Vibrant Varieties

- **'KAJARI' MELON:** An easy-to-grow melon that matures in 70 days, the fruit is a brilliant copper red with green-cream stripes. The pale green flesh is sweet and aromatic.
- **'RICH SWEETNESS 132' MELON:** The fruit is small in size with a vibrant red and yellow striped color. It has an intense sweet aroma and juicy, honeydew-like flesh that matures in 80 days.
- **'TIGGER' MELON:** Maturing in 90 days, the fruit is vibrant yellow with brilliant fire-red zigzag stripes. It is the most fragrant and is rich in flavor.
- **'JUBILEE' WATERMELON:** This large, oval-shaped fruit has very sweet, deep red flesh and matures 90 days from seeding.
- **'ROYAL GOLDEN' WATERMELON:** This heirloom has brilliant golden-yellow skin when ripe. It has very tasty pinkish-red flesh that is sweet with a crisp texture. It matures in 80 to 85 days.

RAINBOW BELL PEPPERS

Rainbow bell peppers are among the most visually appealing vegetables commonly cultivated in home gardens. Bell peppers, characterized by their vibrant colors, offer an enjoyable experience for gardeners. As they mature, these sweet bell peppers undergo a captivating transformation, displaying a rainbow of hues. Initially the fruit is green, then they will gradually change color as they ripen, transforming (according to their variety) into red, yellow, orange, white, purple, or lilac colors. Of course, some varieties do retain their green color even when fully ripe.

Planting

Bell pepper plants are annuals and belong to the nightshade family. Bell peppers need a long growing season and to achieve this, seeds are often started indoors, approximately 6 to 8 weeks before the last spring frost date.

Bell peppers thrive in full sun conditions in well-drained, fertile soil. They can be planted directly in the ground, in raised beds, or in containers. Both starting seeds indoors and planting peppers from transplants are viable methods for growing your plants.

As they mature, these sweet bell peppers undergo a captivating transformation, displaying a rainbow of hues.

Sowing Indoors

Sow bell pepper seeds indoors about 6 to 8 weeks before the last frost date. Fill the seedling tray with a seed-starting mix and sow 1 to 2 seeds per cell, ensuring the seeds are at a depth of approximately ½ inch (1 cm). Maintain an adequate and even soil moisture level during the germination phase. Additionally, position the seedling tray in a location with an air temperature from 75°F to 85°F (24°C–29°C). If necessary, use a heat mat to help seed germination. Once the seeds have germinated, remove the heat mat and provide them with indirect sunlight to prevent leggy seedlings. The seedlings are ready for transplanting when they have developed a couple of sets of their true leaves.

Transplants

Transplant seedlings outdoors once the soil temperature reaches 65°F (18°C). Before planting in the garden, acclimate the seedlings slowly to outdoor conditions a little each day for a period of approximately 1 week. When planting them in the garden, space plants 18 to 24 inches (46–61 cm) apart, setting them in the soil deep enough so that their root ball is covered.

Care

Water

Bell peppers require deep watering with approximately 1 to 2 inches (2.5–5 cm) of rain or irrigation per week. During the hottest months, increase the frequency of watering to 2 to 3 times per week.

If you reside in an extremely hot climate, consider using shade cloth to provide protection from the sun for your pepper plants. Additionally, apply organic mulch around the plants at planting time to retain soil moisture, suppress weeds, and regulate soil temperature.

↑ A fully colored pepper ready for harvest

↑ A summer harvest of peppers and tomatoes

Fertilizer

Bell peppers are heavy feeders and require regular fertilization. Apply fertilizer with a high phosphorus content at planting time and again when the plant is blooming. This will encourage fruiting. Avoid excessive nitrogen, which can cause excessive leaf growth at the expense of fruit production.

Pruning and Support

Bell peppers can grow quite tall, necessitating support to prevent them from toppling over. This can be achieved by using stakes or tomato cages. While pruning is not mandatory, it is recommended to remove any branches that do not have any flower buds, redirecting the plant's energy toward the development of existing peppers.

Companion Plants

Bell peppers grow well with basil, carrots, eggplants, nasturtiums, marigolds, onions, chives, cilantro, dill, oregano, spinach, and tomatoes. Avoid growing bell peppers with beans, brassicas, and fennel.

Pests and Diseases

Bell peppers are susceptible to a range of pests and diseases. The most prevalent pests include flea beetles, slugs, snails, aphids, spider mites, cutworms, and caterpillars. Additionally, it is prone to diseases such as leaf spot. (See page 150.)

Harvest and Storage

When to Harvest

In general, the different colors of bell peppers tend to mature at a slightly slower pace compared to the green bell pepper varieties.

Bell peppers are ready for harvesting after approximately 90 days from seeding, when they have reached their full fruit size. The color of the pepper will serve as an indicator of the optimal time to pick them.

Bell peppers have delicate stems, so it is crucial to use sharp pruners or scissors to cut the fruit from the plants. Pulling the fruit off can damage the stem and wound the plant, leading to a site for bacteria to enter.

How to Store

Rainbow bell peppers are at their peak flavor when freshly harvested and can be enjoyed in various culinary preparations, including eaten raw, baked, grilled, or stir-fried. Should you have an abundance of harvest, you can store them at room temperature for a few days or refrigerate them for up to 2 weeks.

For extended storage, consider freezing peppers. Remove the stems and seeds, and slice or chop them into small pieces. Arrange the pieces in a single layer on a baking sheet and freeze them. Once frozen, transfer them to a sealable, plastic freezer bag. Frozen peppers can maintain their quality for up to 8 months.

My Favorite Vibrant Varieties

- **'CHOCOLATE BEAUTY' BELL PEPPER:** A big, beautiful bell pepper with a chocolate color offering a burst of sweetness, this variety matures 75 to 80 days after seeding.
- **LILAC BELL PEPPER:** This sweet bell pepper matures in 70 to 80 days and has a light purple color.
- **PURPLE, ORANGE, RED, AND YELLOW BELL PEPPERS:** These sweet bell peppers have thick flesh in brilliant rainbow colors that is nicely flavorful, maturing in 75 days.
- **RAINBOW BLEND BELL PEPPER:** An assortment containing a full range of mature fruit colors in red, orange, yellow, chocolate, lilac, ivory, and green. They mature in 60 days.
- **'VIOLET SPARKLE' BELL PEPPER:** This bell pepper has a pointy shaped fruit that is purple streaked with pale yellow. It's sweet and delicious, maturing in 75 days from seed.

HOW TO OVERWINTER PEPPER PLANTS

↑ A variegated pepper in the garden

Peppers are typically grown as annuals, but in their native environment, pepper plants are perennials. They can survive for several years under the right conditions. To extend the growing season or avoid the inconvenience of starting and replanting new pepper plants in the spring, overwintering can be a viable option.

Pepper plants can be overwintered indoors. But if you reside in a temperate climate with minimal frost, outdoor overwintering is also possible.

Here are some tips to overwinter pepper plants indoors.

1. **Step 1.** Harvest any remaining fruits and prune the plants thoroughly.
2. **Step 2.** Dig up the pepper plant and shake off the soil adhering to the roots. Trim the roots back and gently wash the soil off the roots to prevent the spread of diseases.
3. **Step 3.** Fill a small pot with potting mix and plant the pepper plant into the fresh soil. Gently press the soil down around the plant.
4. **Step 4.** Place the pepper plant in a cool location with temperatures between 50°F and 65°F (10°C–18°C) and where it will receive moderate amounts of sunlight. Water thoroughly and then check it regularly for watering, keeping the soil evenly moist.

Transplant your pepper plants back outdoors at least 3 to 4 weeks before the anticipated last spring frost date. Watch for any surprise freezing temperature drops and cover if necessary for protection.

RAINBOW HOT PEPPERS

Hot peppers, also known as chili peppers, exhibit a diverse range of shapes, colors, and numerous varieties. The heat level of hot peppers is directly proportional to the number of blister-like sacs filled with capsaicinoids that are present on the interior wall of the peppers. The main chemical is capsaicin, a compound found in the white pith of a pepper's inner wall where the seeds are attached. Capsaicin is responsible for the spicy heat sensation when eating or handling hot peppers. The Scoville scale is a measurement system used to quantify capsaicin content in peppers—indicating the level of heat or spiciness. Peppers with a higher Scoville heat unit (SHU) are hotter than those with a lower rating. For instance, while sweet peppers have no SHU rating, jalapeño peppers have a range of 2,500 to 8,000 SHU. Did you know that the world's hottest pepper that is commercially available is the Carolina Reaper, with a heat rating of 2.2 million SHU?

Planting

Rainbow hot pepper plants are annuals and belong to the nightshade family. Hot peppers have a long growing season. Seeds are often started indoors, approximately 6 to 8 weeks before the last spring frost date.

Hot peppers thrive in full sun and like well-drained, fertile soil for optimal growth. They can be planted directly in the ground, in raised beds, or in containers. Both starting seeds indoors and planting from transplants are viable methods for growing hot peppers.

↑ Hot peppers come in a broad array of shapes and colors

↑ Leave peppers on the plant so they can develop their full color

Sowing Indoors

Sow hot pepper seeds indoors about 6 to 8 weeks before the last frost date. Fill the seedling tray with a seed-starting mix and sow 1 to 2 seeds per cell, ensuring each are at a depth of approximately ½ inch (1 cm). Maintain adequate soil moisture during the germination phase. Additionally, position the seedling tray in a location with air temperatures between 75°F and 85°F (24°C–29°C). If necessary, use a heat mat to help seed germination.

Once the seeds have germinated, remove the heat mat and provide them with indirect sunlight to prevent leggy seedlings. The seedlings are ready for transplanting when they have developed a couple of sets of their true leaves.

Transplants

Transplant hot pepper seedlings outdoors once the soil temperature reaches 65°F (18°C). Before moving them permanently outdoors, acclimate the seedlings gradually to outdoor conditions over a period of approximately one week. Space plants from 18 to 24 inches (46–61 cm) apart, planting them in the soil deep enough so that their root ball is covered.

Care

Water

Hot peppers require consistent watering, and it is crucial to main-

tain a good soil moisture level throughout the summer season. Irrigate hot peppers with 1 to 2 inches (2.5–5 cm) of water per week. During the hottest months, increase the frequency of watering to 2 to 3 times per week. If you reside in an extremely hot climate, consider using shade cloth to provide protection for your pepper plants. After planting, apply organic mulch around the plants to retain soil moisture, suppress weeds, and regulate soil temperature.

Fertilizer

Hot peppers require a balanced fertilizer with a high phosphorus content at planting time and again when the plant sets flowers. This will encourage fruiting. Avoid high-nitrogen fertilizers, which can cause excessive leaf growth instead of fruits. It is crucial to avoid overfertilizing hot pepper plants, as this can delay flowering and fruiting.

Pruning and Support

Hot peppers can grow quite tall, requiring them to have support to prevent them from toppling over. This can be achieved by using stakes or tomato cages. Pinch off any flowers before the plant reaches its mature size. Also prune out some of the branches that have not produced flower buds. This will redirect the plant's energy elsewhere to other branches for the development of peppers.

Companion Plants

Hot peppers grow well with basil, carrots, eggplants, nasturtiums, marigolds, onions, chives, cilantro, dill, oregano, spinach, and tomatoes. Avoid growing hot peppers with beans, brassicas, and fennel.

Pests and Diseases

Hot peppers are susceptible to flea beetles, slugs, snails, aphids, spider mites, cutworms, and caterpillars. Additionally, hot peppers are prone to diseases such as leaf spot. (See page 150.)

Harvest and Storage

When to Harvest

Hot peppers are ready for harvesting from midsummer to early fall. Due to their delicate stems, it is imperative to use sharp pruners or scissors to cut the fruits from the plants. Pulling the fruit can damage the stem, leaving a spot to introduce harmful bacteria into the plant. Additionally, wearing gloves during harvesting can help prevent capsaicin from coming into contact with your skin.

How to Store

Hot peppers can be stored in the refrigerator for up to 2 weeks. For extended storage, consider freezing the peppers. Frozen peppers can maintain their quality for up to 8 months. You can also dry or pickle peppers or make chili oil with your harvest.

My Favorite Vibrant Varieties

- **'BUENA MULATA' PEPPER:** This stunning purple hot pepper matures in 75 to 85 days.
- **'CHOCOLATE 7 POT' PEPPER:** Maturing within 90 to 120 days from seeding, this supremely spicy pepper has a dark chocolate color.

↑ My hot pepper harvest—look at all the colors!

- **'DEATH SPIRAL' PEPPER:** This pepper with a twisted shape ranks among the world's hottest peppers and matures in 90 to 120 days.
- **'GORONONG' PEPPER:** This rare habanero-type pepper from Malaysia has beautifully contorted fruit in a canary yellow color. It matures 90 to 100 days from seeding.
- **'PUMA' PEPPER:** Each fruit has brushstrokes of tangerine and violet colors with a habanero-level heat rating. It's ready for harvesting 90 to 100 days from seeding.

COLORFUL WINTER SQUASH

Winter squash is distinguished from summer squash by its thick skin, denser flesh, sweetness, and storability. The majority of winter squash belong to the *Cucurbita maxima* species and exhibit a diverse range of colors, shapes, and sizes. The most widely consumed winter squash varieties include pumpkins and butternut, acorn, and spaghetti squash. Winter squash are renowned for their nutritional value, providing an abundance of vitamin A, vitamin C, and potassium.

Winter squash is distinguished from summer squash by its thick skin, denser flesh, sweetness, and storability.

Planting

Winter squash requires a long growing season and will not tolerate cold weather. They thrive in warm conditions and exhibit moderate drought tolerance once established. Winter squash grow best in full sun with fertile, moist, well-drained, loose, and sandy soil conditions. They prefer a slightly acidic to neutral pH soil.

Sowing Indoors
Winter squash are easy to start from seeds. Sow the seeds indoors 3 to 4 weeks before transplanting them outdoors. Fill a seedling tray with a seed-starting mix and sow 1 seed per cell, ensuring a depth of approximately 1 inch (2.5 cm). Maintain adequate soil moisture during the germination phase. Place the seedling tray where temperatures are between 65°F and 75°F (18°C–24°C). If necessary, use a heat mat to help seed germination. Once the seeds have germinated, remove the heat mat and provide them with indirect sunlight to prevent leggy seedlings. Seedlings are ready for transplanting once they have developed a few sets of their true leaves. Germination typically occurs within 5 to 15 days,

↑ Squash seeds can be started indoors if you have a short growing season.

depending on the variety and soil temperature.

Care

Water

Winter squash needs a lot of water, so keep the soil moist—but not waterlogged. Winter squash plants have large leaves that tend to wilt in the hot afternoon sun, but they will often promptly recover upon watering, or when the sun goes down.

Fertilizer

Fertilize the plants with a well-balanced fertilizer once they have commenced flowering.

Pollination

Winter squash plants produce both male and female flowers on the same plant. Certain varieties of winter squash are self-pollinating, while others need help with manual hand pollination to produce a bountiful harvest of fruit. If you observe that the female flowers are falling off, this indicates that they have not been adequately pollinated. In this situation, hand pollination can be employed to help. While hand pollination can be time-consuming, it is an effective method for achieving a substantial crop yield.

The most important aspect of hand pollination is to distinguish between the male and female flowers—which both develop on the same plant. Male flowers have a short stem, while female flowers bear a small fruit at their base. When flowers first open in the early morning, take a small paintbrush or cotton swab to collect pollen from the male flower. Then, gently transfer this pollen to the center of the female flower.

Pruning and Support

Certain winter squash varieties produce tendrils and exhibit a vinelike growth habit. Consequently, it is imperative to use a sturdy trellis to support their growth. This practice not only maximizes space but also keeps the squash off the ground.

As the plant grows, prune the leaves that touch the ground. When pruning, make clean cuts close to the main stem. Regular pruning also ensures adequate airflow, which helps prevent fungal diseases like powdery mildew.

Companion Plants

Winter squash grow well with corn, sunflowers, peas, beans, borage, chamomile, nasturtiums, marigolds, dill, oregano, parsley, chives, calendula, lemon balm, and peppermint. Avoid growing winter squash with potatoes, beets, fennel, and melons.

Pests and Diseases

The most common pests and diseases of winter squash are beetles, squash bugs, squash

↑ A female squash flower

↑ A male squash flower

↑ Red 'Kuri' squash

vine borer, and powdery mildew. (See page 150.)

Harvest and Storage

When to Harvest

Winter squash is typically ready for harvesting 85 to 120 days after seeding or planting. In contrast to summer squash, winter squash requires a longer period on the vine to fully mature. Some gardeners just leave them attached to the vine until the first frost. Winter squash has a thick protective skin and can tolerate the longer time in the garden. Its flesh will still be dense and firm on the inside.

How to Store

When fully ripe, winter squash will retain its shelf life more effectively. After harvesting, leave the fruit in direct sunlight outdoors or in a greenhouse for 1 week. This curing process hardens the outer skin, facilitating optimal storage. Next, bring the fruit indoors and store it in a cool, dry, and well-ventilated location.

My Favorite Vibrant Varieties

- **'DELICATA' SQUASH:** This winter squash has a high sugar content and a rusted white or yellow skin color with green stripes. It takes 100 days to reach maturity from seed.
- **'ESSEX HYBRID' SQUASH:** This large warty-looking winter squash is a cross between a turban and hubbard squash. It has a rich and sweet flavor and matures in 100 days from seeding.
- **'HONEY BUN' SQUASH:** This eye-catching winter squash has a variegated round shape. The fruit turns a caramel bronze color when ripe and matures in 95 days.
- **'HONEYNUT' BUTTERNUT SQUASH:** This variety yields individual-size butternut squash with a sugary sweet taste and deep orange flesh. It matures in 100 days from seeding.
- **'KAKAI' PUMPKIN:** This yellow-orange pumpkin, mottled here and there with dark green, has large, hull-less seeds that are best for roasting. It matures in 100 days from seeding.
- **'LONGUE DE NICE' SQUASH:** This winter squash can grow from 10 to 20 pounds (5–9 kg). Fruits have a long neck and few seeds. It matures to have salmon-colored flesh about 100 days from seeding.
- **'NORTH GEORGIA CANDY ROASTER' SQUASH:** This winter squash has pink, banana-shaped fruit and a blue tip. The flesh is orange. Matures in 95 days with weights of around 10 pounds (5 kg).

RAINBOW TOMATOES

Cultivating tomatoes is a rewarding experience for home gardeners due to their popularity, ease of cultivation, and high yield. The exceptional flavor of homegrown tomatoes, characterized by a sweetness and freshness unmatched by commercially produced varieties, is a significant draw. My personal experience underscores this. I previously disliked tomatoes, but the intense burst of flavor from my first homegrown tomato was transformative and unforgettable. Tomatoes have a lot of nutritional benefits, including potassium, iron, vitamin C, and other antioxidants.

Planting

Tomatoes are long-season, heat-loving vegetables that cannot tolerate frost. There are hundreds of tomatoes varieties in many different colors, shapes, and flavors.

There are two main types of tomatoes:

Determinate Tomatoes: This type is also known as bush tomatoes. These tomato varieties are compact and mature early, reaching a height of 2 to 4 feet (61–122 cm). They require minimal pruning. Determinate tomatoes produce fruit all at one time and are an excellent choice for containers as they do not need staking or trellis support.

Indeterminate Tomatoes: These tomatoes are my preferred choice for cultivation, and they are an excellent type for most gardeners. Indeterminate varieties will continue to grow indefinitely as long as the weather remains warm, resulting in higher yields than those of the determinate types. Indeterminate tomato plants grow as vines and consequently require support from stakes, cages, or trellises.

Sowing Indoors
Start tomato seeds indoors approximately 8 weeks before you plan to move them outside into the garden. Fill the seedling tray with seed-starting mix and sow 1 to 2 seeds per cell at a ½ inch (1 cm) depth. Keep the soil evenly moist during the germination period and keep the seedling tray at a room temperature of between 60°F and 70°F (16°C–21°C), otherwise you will need a heat mat for the seeds to help with germination. Remove the heat mat after the seeds have germinated and place the seed tray under a grow light to prevent legginess in the seed-

↑ Indeterminate tomato vines growing on a trellis arch

↑ Determinate tomatoes growing in a container

lings. Tomato seedlings are ready to transplant when they grow a couple sets of their true leaves. Gradually harden them off for about 1 week so they can acclimate to the outdoor conditions.

Transplants

When transplanting tomatoes, it is crucial to plant them at a deeper depth than they were previously growing so they will form many new roots along the lower stem. To do this, remove a few of the bottom leaves. You have a choice of two ways to plant tomato transplants in the garden. The first method is to dig a deep hole and set the root ball in the bottom of the hole so that the first set of leaves is just above the surface of the soil. Alternatively, you can plant the tomato so that it is on its side, with some of the stem buried beneath the soil. When planting using either one of these methods, new adventitious roots will grow along the tomato stem, promoting more root development and enhancing the overall strength of the root system.

↑ The "adventitious roots" growing on the stem of a tomato plant are responsible for enhancing its stability and facilitating the uptake of essential nutrients.

For determinate varieties, space the plants 18 to 24 inches (46–61 cm) apart. For indeterminate varieties, space the plants

WHAT IS BLOSSOM END ROT?

Blossom end rot is a common issue that affects some of the most popular garden vegetables such as tomatoes, eggplant, peppers, and squashes. Blossom end rot is a physiological disorder that manifests in the early stages of fruit development, characterized by a dark stain or water-soaked spot on the blossom end. If left untreated, the stain will progressively enlarge and turn brown. This condition can be attributed to insufficient nutrients (Calcium deficiency), inconsistent watering, and rapid plant growth. The most effective strategy involves incorporating calcium into the soil before planting, along with amendments such as lime or bone meal. Consistent watering throughout the growing season is also crucial.

24 to 36 inches (61–91 cm) apart. Proper spacing is crucial for adequate air circulation, sunlight exposure, and optimal growth.

Care

Water

Consistently water tomato plants to maintain even soil moisture level. Tomato plants require approximately 1 to 2 inches (2.5–5 cm) of water per week. Water deeply at the base of the plant to promote deep root development and prevent fungal diseases.

Fertilizer

Fertilize tomato plants with a well-balanced fertilizer every 3 to 5 weeks. To prevent blossom end rot, supplement the fertilizer with calcium-rich sources such as commercial tomato blossom end rot prevention products, crushed eggshells, bonemeal, or lime.

Pruning

Pruning tomatoes can help the plant to grow stronger and produce healthier fruits. However, not all gardeners prune their tomatoes. Here are a few reasons why you should prune your tomato plants:

- **Better Airflow:** Improved airflow will decrease humidity levels and mitigate the risk of fungal diseases. Additionally, good airflow helps pollinators—thereby enhancing the pollination of blossoms.
- **Bigger Fruits:** By reducing the number of branches and leaves, you can redirect the plant energy toward producing larger fruits. Pruning also facilitates the ripening of fruits, particularly in regions with shorter growing seasons.
- **Control Growth:** If you are growing indeterminate tomato varieties, it is advisable to prune them back. Indeterminate tomatoes will continue to grow until the growing conditions are no longer favorable. Pruning will provide you with control over the direction of their growth and their size.

First, for the health of your plant, cut off the leaves that touch the soil. Then, look at the main stem to see the arrangement of the leaves and branches. The new shoots between the leaves and the main stem are called "suckers." Each sucker will ultimately grow into a side stem that will produce more fruits. Personally, I do not prune suckers on my cherry tomato varieties so I can have a higher yield. Tomato suckers can be easily removed by snapping them off when they are still small. If they are larger, use a sharp, clean pruner to avoid damaging the plant. Some gardeners remove their tomato suckers to produce fewer, but larger tomatoes and a sturdier plant. Others just prune their indeterminate tomatoes late in the season to encourage the more mature fruit to ripen before the frost.

Staking and Support

Regardless of whether you plant determinate or indeterminate tomato varieties, it is essential to provide support to keep plants upright and to prevent the leaves from touching the soil. Install a sturdy stake, cage, or trellis next

HOW TO ROOT TOMATO SUCKERS

Instead of discarding any removed tomato suckers, you can plant them and grow them into new plants. This process is relatively straightforward.

- **Step 1.** Simply root the cuttings in a Mason jar filled approximately 3 inches (7.5 cm) deep with water. Submerge the cut ends immediately in the water after severing them from the mother plant. You can root several cuttings in each jar. Ensure that there are no leaves in the water.
- **Step 2.** Place the jar in a warm location, away from direct sunlight. Change the water every couple of days. In approximately 2 to 3 weeks, you will see roots growing out of the submerged stems.
- **Step 3.** Plant your rooted cuttings in an organic potting mix at the same depth they were submerged in water. Gradually acclimate the plants to outdoor conditions for 1 week before transplanting them outdoors.

↑ Big rainbow tomatoes

to the tomato plant at the time of planting. Secure the stems to the support using plant ties. Continue to secure the stems as the plant grows taller.

Companion Plants

Tomatoes grow well with basil, borage, chives, marigolds, carrots, garlic, lettuce, onions, parsley, celery, nasturtiums, calendula, cilantro, lavender, dill, radishes, sage, squash, and beans. Avoid growing tomatoes with corn, potatoes, brassicas, rosemary, and fennel.

Pests and Diseases

Tomatoes are susceptible to hornworms, early blight, late blight, and leaf spots. (See page 152.)

Harvest and Storage

When to Harvest

Tomatoes typically ripen over a period of 60 to more than 100 days, commencing from midsummer onward. The ripening time can vary depending on the specific tomato variety, climatic conditions, growing practices, and the size of the fruit. Generally, smaller cherry tomatoes tend to ripen faster compared to the larger varieties.

Harvest tomatoes when they reach a firm texture and exhibit a deep red color, regardless of their size. Additionally, harvest other colored tomatoes such as yellow, purple, and rainbow tomatoes once they attain their appropriate color.

How to Store

For short-term storage, ripe tomatoes should be kept at room temperature with the stem side down and away from direct sunlight. If you intend to store tomatoes for a few days, wash and dry your ripe tomatoes, then store them in the refrigerator. However, refrigeration can diminish their flavor and cause a mushy texture.

If you have an excess of ripe tomatoes, you can consider canning, freezing, or drying to preserve them for 12 months or more.

How to Ripen Green Tomatoes

Although vine-ripened tomatoes may be sweeter, if you still have tomatoes on the vine when cooler weather arrives, pick them and place them in a single layer in a cardboard box or paper bag. This will allow them to naturally ripen.

My Favorite Vibrant Varieties

- **'BIG RAINBOW' TOMATO:** This indeterminate type matures in 85 days. It has huge yellow fruit with neon-red streaking that is delicious and sweet,
- **'BLACK SEA MAN' TOMATO:** This determinate tomato matures 75 days from seeding. It has big purple-black fruit with a rich flavor.
- **'BLUE BEAUTY' TOMATO:** This indeterminate beefsteak-type tomato matures in 80 days and is a cross between 'Beauty King' and a blue tomato.
- **'INDIGO BLUE CHOCOLATE' TOMATO:** This indeterminate type matures 80 days from seeding; it is sweet and juicy with black anthocyanin-rich skin.
- **'QUEEN OF THE NIGHT' TOMATO:** This indeterminate type matures in 80 days with medium-size fruit that ripen to have brushstrokes of ebony, crimson, and orange colors.
- **'SART ROLOISE' TOMATO:** This indeterminate tomato has a sweet and fruity taste, maturing 80 to 90 days after seeding. It has gorgeous globe-shaped fruit with stunning purple and yellow colors.
- **'TASMANIAN CHOCOLATE' TOMATO:** This determinate type matures 70 days after seeding. It is a dwarf cultivar that yields red beefsteak-type fruits.
- **'ZEBRA' CHERRY TOMATO:** This determinate tomato matures in 70 days with a medium-size cherry fruit that is burgundy red and green striped.

GREEN, YELLOW, AND BICOLOR ZUCCHINI

Zucchini, a variety of summer squash belonging to botanical plant group *Cucurbita pepo*, is a highly sought-after garden vegetable. This broad group encompasses a range of summer squash, including green and yellow zucchini, bicolor zucchini, pattypan squash, and many others. Not only is zucchini relatively easy to cultivate, but it is also both delectable and nutritious. The culinary possibilities for zucchini are boundless, featuring appearances in pasta, salads, bread, cakes, casseroles, pizza, and soups.

Planting

Zucchini is a warm-season vegetable that flourishes in full sun with well-drained, fertile soil. The plant typically takes approximately 45 to 55 days to initiate flowering—then watch out, here come the harvests. Zucchini can be direct-sown from seed in the garden or started indoors.

Direct-Sowing

Direct-sow zucchini seeds after the last spring frost has passed and the soil temperature has reached at least 60°F (16°C). Sow zucchini seeds 1 inch (2.5 cm) deep and approximately 24 inches (61 cm) apart. Maintain even soil moisture until the seeds sprout; usually this is within 10 to 14 days.

Sowing Indoors

Start the seeds indoors 3 to 4 weeks before transplanting the plants outdoors. Fill the seedling tray with a seed-starting mix and sow 1 seed per cell, ensuring a depth of approximately 1 inch (2.5 cm). Maintain adequate soil moisture during the germination phase. Additionally, position the seedling tray in a location with an air temperature between 65°F and 75°F (18°C–24°C). If necessary, use a heat mat to help seed germination. Germination typically occurs within 5 to 15 days, depending on the variety and soil temperature. Once the seeds have germinated, remove the heat mat and provide the seedlings with indirect sunlight to prevent legginess. Seedlings are ready for transplanting once they have developed a few sets of their true leaves.

Care

Water

Zucchini plants require a lot of water. Ensure that the soil remains consistently moist but not waterlogged. Water the plants when the top layer of soil feels dry to the touch. The large, broad leaves of zucchini plants are prone to wilting in the afternoon sun, but they promptly recover with watering. Inconsistent watering can result in irregularly shaped fruits.

Fertilizer

Fertilize the plants with a well-balanced fertilizer once they have started to flower.

Pollination

Zucchini plants produce both male and female flowers on the same plant. If you notice that the female flowers are falling off, this indicates that they have not been adequately pollinated. In such a situation, hand pollination can be employed to help this process. While hand pollination can be time-consuming, it is an effective method for achieving a substantial crop yield.

Male flowers have a short stem, while female flowers bear a small fruit at their base. When the new flowers open in the early morning, use a small paintbrush or cotton swab to collect pollen from the male flower. Then, gently transfer this pollen to the center of the female flower to pollinate the flower. Continue until all the female flowers are pollinated. Repeat the next morning.

Pruning and Support

Trellising vining varieties keeps the zucchini off the ground, thereby facilitating the prevention of diseases.

As the plant grows, prune the leaves that touch the soil. When pruning, make clean cuts close to the main stem to avoid leaving stubs. Regular pruning helps prevent fungal diseases like powdery mildew.

Companion Plants

Zucchini grow well with corn, sunflowers, peas, beans, borage, chamomile, nasturtiums, marigolds, dill, oregano, parsley, chives, calendula, lemon balm, and peppermint. Avoid growing zucchini with potatoes, beets, fennel, and melons.

Pests and Diseases

The most common pests and diseases of zucchini are beetles, squash bugs, squash vine borers, and powdery mildew. (See page 150.)

Harvest and Storage

When to Harvest

Zucchini are ready for harvesting when the fruits are approximately 4 to 6 inches (10–15 cm) long and firm to the touch. Young zucchini is tender and flavorful. Regularly inspect zucchini plants, as they grow rapidly. If left unharvested, zucchini can get quite large and develop hard seeds and rinds.

Use a sharp knife or sanitized pruner to remove the fruit, leaving approximately 1 inch (2.5 cm) of the stem attached to the top of the fruit.

How to Store

To preserve zucchini, store the whole, unwashed fruit in the refrigerator. This storage method will help maintain its freshness for approximately two weeks.

For extended storage, you can freeze the zucchini. Wash them, then cut them into the desired pieces, blanch them in boiling water for 1 to 2 minutes, pat them dry, spread them out on a baking sheet, and freeze them overnight. Once frozen, store them in a sealed plastic bag or an airtight container and keep them in the freezer for up to 6 months.

↑ A male zucchini flower

↑ A female zucchini flower

My Favorite Vibrant Varieties

- **'COSTATA ROMANESCO' ZUCCHINI:** This Italian heirloom has tender, gray-green skin and prominent ribbing. It matures in 54 days.
- **GOLDEN ZUCCHINI:** This summer squash has delicious, slender fruit that is bright golden-yellow. It is a bush plant that matures in 54 days.
- **'SUNBURST HYBRID' PATTYPAN SQUASH:** A delightful, yellow, scalloped fruit that has marked green at both ends, this squash is sweet with a tender texture and matures in 50 days.
- **'ZEPHYR' SQUASH:** The slender fruits are yellow with faint white stripes and a light green blossom end. This squash is delicious, with a nutty flavor and firm texture. It matures in 54 days.

Seed and Pod Vegetables

PURPLE AND YELLOW SNAP PEAS

Purple and yellow snap peas have a strikingly vibrant color on their outer pods and feature neon green peas within. Snap peas are a productive and easy-to-grow vegetable. They are the hybrid offspring of shell peas and snow peas, so it is crucial to distinguish between snap peas and snow peas. Snap peas possess plump pods, while snow peas have flat and bendable pods.

Snap peas are a traditional cool-season vegetable commonly cultivated in home gardens. Growing colorful snap peas enhances visual appeal and brings a sense of joy to the garden, and also provides a rewarding and nutritious snack. Both the sweet pod and the pea are edible! In my garden, snap peas rarely make it indoors.

Planting

Snap peas, an annual vegetable, are a cool-season crop. Timing is crucial when planting snap peas. If they are planted too early, they can be stunted by soil that is too cold, or frosty temperatures. For spring planting, sow the seeds as soon as the soil becomes workable, which is typically around 6 weeks before the last expected spring frost date. For fall planting, sow the seeds in late summer, approximately 8 to 10 weeks before the first frost date.

Direct-Sowing

Snap peas flourish in full sun when they receive at least 6 hours of direct sunlight daily. They thrive in fertile and well-drained soil with a slightly acidic to neutral pH. Sowing snap peas directly from seed is the preferred method to grow them, as they do not tolerate transplanting well.

Before sowing, it is advisable to soak the seeds overnight to

COLORFUL SNOW PEAS

If you like snow peas, there are colorful varieties to add to your garden. Try the 'Golden Sweet' Snow Pea, which features large, golden-yellow pods and purple flowers. It matures 61 days from seeding.

↑ Yellow snap peas

Growing colorful snap peas enhances visual appeal and brings a sense of joy to the garden, and also provides a rewarding and nutritious snack.

enhance germination. Plant the seeds approximately 1 inch (2.5 cm) deep and 3 inches (7.5 cm) apart, with rows spaced 24 to 36 inches (61–91 cm) apart. Keep the soil evenly moist. Snap pea seeds typically germinate in 10 to 14 days.

Care

Water

Snap peas require consistent watering, particularly during the germination phase. It is crucial to avoid allowing the soil to dry out between waterings. Snap peas typically need approximately 1 inch (2.5 cm) of water per week. To ensure proper watering, direct the water at the base of the plant rather than onto the leaves to help prevent the spread of fungal diseases.

Fertilizer

Snap peas are naturally a nitrogen-fixing crop, so they may not require a significant amount of additional nitrogen fertilizer. If you do wish to give them a little boost, you can add an all-purpose, granular fertilizer and fish meal fertilizer into the soil at planting. Follow the label directions.

Support

The majority of snap pea varieties are vining, which means that they will produce tendrils to help themselves climb. These tendrils

can be used to guide the growth of the plants along a support such as a wire fence, trellis, twine, stake, netting, or bamboo tripod. Tall vines will start to climb quickly after sprouting, so make sure to set up the support at planting time.

Companion Plants

Beneficial companion plants for snap peas include cilantro, mint, radishes, lettuce, spinach, corn, and cucumbers. Avoid growing snap peas with onions, garlic, leeks, and shallots.

Pests and Diseases

The most common pests and diseases for snap peas are aphids, slugs, snails, root rot, and powdery mildew. (See page 150.)

Harvest and Storage

When to Harvest

Snap peas can be harvested at various stages throughout the growing season. They are ready for harvesting when the pods begin to swell, but they should not be left on the plant to get overly large. You can test the readiness of snap peas by snapping a pod. If it produces a clear snapping sound, it is ready. Another method is to gently squeeze the pod; if the peas inside feel plump, they are ready. Regularly harvest snap peas to encourage the plant to keep on producing more.

Both the pods and the peas within them are edible, making snap peas an ideal choice for snacking and stir-frying.

↑ Purple snap peas

↑ Pea seeds ready for planting

To harvest snap peas, use pruning shears or scissors to cleanly cut the pod from the vine, thereby preventing any potential damage to the plant.

How to Store

Snap peas are most flavorful when consumed immediately after harvesting. However, if you have an excess harvest, you can try the following methods to prolong their shelf life. Keep snap peas unwashed in a plastic bag in the crisper drawer of your refrigerator for up to 10 days. Blanch snap peas in boiling water for 1 minute. Immediately transfer them to an ice bath to stop the cooking process. Then, drain the peas and place them in a sealable plastic freezer bag. Store them in the freezer for up to a year.

My Favorite Vibrant Varieties

- **'HONEY SNAP II' SNAP PEA:** You'll love the eye-catching golden-yellow pods on this snap pea 58 days from seeding.
- **'KELVEDON WONDER' SNAP PEA:** This dwarf snap pea variety is perfect for a container. It produces heavy crops of sweet, small pods 75 days after seeding.
- **'KING TUT PURPLE' SNAP PEA:** This unusual snap pea features striking fuchsia purple flowers and plump purple pods. It matures 70 days after seeding.
- **'SUGAR BON' SNAP PEA:** This dwarf variety produces huge yields of crispy, sweet pods and peas 55 days from seeding.
- **'SUGAR MAGNOLIA' SNAP PEA:** This purple-podded sugar snap pea matures in 50 days.

RAINBOW BEANS

Cultivating rainbow bean varieties is a gratifying, enjoyable, and accessible gardening endeavor. Rainbow beans are among my preferred vegetable crops to cultivate during the summer months. They exhibit a diverse array of characteristics, including various colors, sizes, textures, and plant structures.

Rainbow beans share common cultivation requirements with green beans and can be categorized into two primary types: **bush beans** and **pole beans.**

Bush Beans: As their name implies, bush beans have a lower growth habit. They need minimal care and produce yields faster than pole beans. Bush beans typically commence production within 50 to 55 days, while pole bean varieties require approximately 60 to 65 days to mature.

Pole Beans: Pole bean varieties are vining plants that can grow from 6 to 10 feet (183–305 cm) high and require support. This type tends to produce more beans than the bush bean varieties and is generally more disease resistant.

Planting

Rainbow beans thrive in warm climates and are sensitive to frost. Select a location that receives at least 6 to 8 hours of direct sunlight daily. Beans grow best in well-drained soil that has been enriched with organic matter, such as compost. Rainbow beans are best direct-sown in the garden as they are sensitive to root disturbance. Direct-sow seeds 2 to 3 weeks after the last frost date, and when the soil temperature is at least 60°F to 65°F (16°C–18°C). For bush beans, sow seeds approximately 1 to 2 inches (2.5–5 cm) deep and 2 to 4 inches (5–10 cm) apart. For pole beans, sow seeds 1 to 2 inches (2.5–5 cm) deep and 4 to 6 inches (10–15 cm) apart.

Care

Water

Rainbow beans require consistent watering. Allow the soil

↑ Planting bean seeds

surface to dry to a ½ inch (1 cm) depth between waterings. Beans need approximately 1 inch (2.5 cm) of water per week. Water at the base of the plant to prevent fungal diseases on the leaves.

Fertilizer
Rainbow beans are legumes that do not need extensive fertilization as they possess the ability to fix their own nitrogen back into the soil from the air. In the event that plants exhibit pale or stunted growth, a low-nitrogen fertilizer can be applied to stimulate growth.

Support
Pole beans are climbing vines that require support. Install a trellis, stakes, or tripod structure for the beans to climb up.

Companion Plants

Rainbow beans grow well with corn, squash, marigolds, nasturtiums, carrots, cucumbers, radishes, Brassicaceae family members, beets, lettuce, and strawberries. Avoid growing rainbow beans with onions, garlic, chives, leeks, scallions, and shallots.

Pests and Diseases

The most common pests and diseases of beans are cutworms, beetles, aphids, root rot, leaf spots, and powdery mildew. (See page 150.)

Harvest and Storage

When to Harvest
Bush beans are ready to harvest in about 45 to 55 days after sowing. Pole beans, on the other hand, require approximately 55 to 65 days to mature.

For rainbow beans, the optimal time to harvest is when the pods reach a length of 6 to 8 inches (15–20 cm). Early harvesting is recommended to ensure that the beans are sweet and tender—before they fully mature. Consistent harvesting is crucial to stimulate the plant to produce more future pods.

↑ A colorful bean harvest

Fresh beans are always the best for salads, stir-fries, or steaming.

Beans can be manually picked or cut from the plant using scissors. Scissors or pruners will ensure minimal damage to the plant during harvests.

How to Store

Fresh beans are always the best for salads, stir-fries, or steaming. However, if you have an excessive harvest, you can try the following methods to prolong their shelf life. Keep the beans unwashed in a plastic bag in the crisper drawer of your refrigerator for up to 10 days. You can also blanch beans in boiling water for 1 minute. Immediately transfer them to an ice bath to stop the cooking process. Drain the beans and store them in a freezer bag in the freezer for up to a year.

My Favorite Vibrant Varieties

- **'BLAUHILDE' BEAN:** This pole bean has stringless, purple pods that stay tender. The bean matures 65 days from seeding.
- **'BLUE LAKE' BEAN:** This superb bush bean has huge yields of green pods with a great taste and matures in 60 days.
- **'CHEROKEE TRAIL OF TEARS' BEAN:** This pole bean with red pods and a shiny black skin is good as a snap bean or for drying. It matures in 65 days from seeding.
- **'DRAGON TONGUE' BEAN:** This bush bean has pods that are yellow with purple streaks and an amazing flavor. It reaches maturity in 60 days.
- **'GOLDEN BUTTERWAX' BEAN:** This bush bean has an impressive yield of yellow pods and a superior flavor. It matures 50 days after seeding.
- **'NAUTICA' BEAN:** This dwarf bush bean has tender, green, stringless pods and matures 55 to 60 days from seeding.
- **'PURPLE TEEPEE' BEAN:** This bush bean has flavorful and tender purple snap bean pods. It matures 60 days from seeding.

SCARLET RUNNER BEANS

Scarlet runner beans, botanically known as *Phaseolus coccineus*, are a versatile, attractive, and productive plant that is ideal for home gardens. They are appreciated both for their ornamental, vibrant scarlet red flowers and for their colorful, edible beans. The flowers will attract pollinators, including bees and hummingbirds, making them an excellent choice for wildlife-friendly gardens.

The seeds contained within the pods have a captivating array of colors, transforming from lavender, purple, and pink to a purplish-black hue with purple or light brown mottling as they mature. Like other common beans, scarlet runners are a nutrient-dense food source, providing an abundance of vitamin C, vitamin K, and folate. They are also high in fiber and low in carbohydrates and calories.

Planting

Scarlet runner beans exhibit similarities to other bean varieties. They are perennials in mild climates but are grown as annuals in colder regions. They thrive in warm, sunny locations with well-drained soils that are rich in nutrients. For optimal growth, direct-sow scarlet runner beans in the garden. But, if you want early blooms, you can sow the seeds indoors.

Sowing Indoors

Start the seeds indoors a few weeks before the average last frost date in the spring. Fill the seedling tray with a seed-starting mix and sow 1 seed per cell, ensuring that the seed is at a depth of approximately 1 inch (2.5 cm). Maintain adequate and consistent soil moisture during the germination phase. Additionally, position the seedling tray in a location with a room temperature between 75°F and 85°F (24°C–29°C). If necessary, use a heat mat to help seed germination. Once the seeds have germinated, remove the heat mat and provide them with indirect sunlight to prevent leggy seedlings. The seedlings are ready for transplanting when they have developed a couple of sets of their true leaves. Before planting the seedlings outdoors, gradually acclimate the seedlings to outdoor conditions over a period of approximately 1 week.

Direct-Sowing

Sow seeds outdoors in the late spring when the soil temperature is warm and nighttime temperatures consistently reach approximately 65°F (18°C). Before planting, soak the seeds for approximately 12 to 24 hours to help the germination process. Place the seeds 1 to 2 inches (2.5–5 cm) deep and 4 to 8 inches (10–20 cm) apart. Maintain consistent soil moisture during the germination phase. The seeds typically will germinate in 10 to 14 days.

Care

Water

Scarlet runner beans require consistently moist soil, but not waterlogged conditions. Avoid overwatering, which can lead to root rot. Water deeply once or twice a week, particularly during dry periods. Water at the base of the plant to prevent fungal diseases on the leaves. Applying a layer of organic mulch will help in retaining soil moisture and help suppress weeds.

Fertilizer

Use a balanced, low-nitrogen fertilizer to stimulate flowering and bean production.

Avoid excessive nitrogen, as it promotes foliage growth at the

↑ Scarlet runner bean seeds are large and easy to plant

expense of flowers and pods. But if the plants look pale or have stunted growth, a low-nitrogen fertilizer can be applied to stimulate growth.

Pruning and Support

Pinch out the tips of the vines if they are growing too long or are becoming unmanageable. This can encourage bushier growth and promote good air circulation. Place supports such as poles, tripod structures, string lines, or netting near the plants at the time of sowing or transplanting.

Companion Plants

Scarlet runner beans grow well with salvia, zinnia, corn, squash, cucumbers, radishes, carrots, marigolds, nasturtiums, lettuce, spinach, beets, and onions. Avoid planting scarlet runner beans with garlic, leeks, shallots, sunflowers, and peppers.

Pests and Diseases

The most common pests and diseases for scarlet runner beans are aphids, slugs, spider mites, beetles, and powdery mildew. (See page 150.)

Harvest and Storage

When to Harvest

The optimal time to harvest scarlet runner beans is when they are young—similar to harvesting snap beans. The scarlet runner bean pods should be approximately 6 to 8 inches (15–20 cm) in length, and both the pod and the inside beans are edible and flavorful. Continuous harvesting of the beans will encourage the production of new flowers and enhance future yields. Notably, the flowers are also edible and can be incorporated into salads.

If you harvest the pods when the seeds are fully formed, these are then considered shell beans for drying. Allow the pods to mature on the vine until they turn brown and are dry.

How to Store

There are a few options when it comes to preserving your bountiful harvest. Young pods can be prepared similarly to green beans. They can be canned or blanched and stored in the freezer.

To successfully store the mature seeds, they need to be completely dry. Lay the mature pods on a drying rack in a well-ventilated area until the seeds are fully dry and hard. Once dry, shell the seeds and store them in a glass jar in a cool, dark location. The dried beans can subsequently be used in culinary preparations or for future crop cultivation.

My Favorite Vibrant Varieties

- **'BLACK KNIGHT' BEAN:** This vining bean has young, tender pods and scarlet red blooms. It matures 65 to 80 days from seeding.
- **'GOLDEN SUNSHINE' BEAN:** This tall bean has brilliant chartreuse foliage and vibrant red flowers; it matures in 85 days.
- **'HIDATSA RED' BEAN:** This semi-vining bean has a heavy yield of hardy, rose-red beans for drying, and matures 85 from seeding.
- **'SUNSET' BEAN:** This pole bean with peach to pink blossoms and bright pink-purple seeds matures in 65 to 70 days.

They are appreciated both for their ornamental, vibrant scarlet red flowers and for their colorful, edible beans.

MULTICOLORED YARDLONG BEANS

Yardlong beans, also known as asparagus beans, Chinese long beans, or snake beans, are a rewarding summer crop that is both delicious and highly productive. My favorite yardlong variety to grow is the 'Red Noodle', which I have found to be exceptionally flavorful. This variety is visually striking, with its vibrant reddish-purple blossoms resembling those of sweet pea flowers.

Planting

Yardlong beans are heat-loving vines that thrive in warm climates and are very sensitive to frost. Select a location that receives at least 6 to 8 hours of direct sunlight daily. Yardlong beans grow best in well-drained soil that has been enriched with organic matter, such as compost. Growing yardlong beans is quite similar to growing pole beans. They prefer direct-sowing and dislike having their roots disturbed.

Sow seeds in late spring when the soil temperature is warm and nighttime air temperatures consistently reach approximately 55°F (13°C). Direct-sow the seeds 1 to 2 inches (2.5–5 cm) deep and 4 to 6 inches (10–15 cm) apart. Maintain even soil moisture during the germination phase.

Care

Water
Yardlong beans need consistent watering. Allow the soil surface to dry out to approximately ½ inch (1 cm) between waterings. These beans require approximately 1 inch (2.5 cm) of water per week, and the amount of water should be increased during

This variety is visually striking, with its vibrant reddish-purple blossoms resembling those of sweet pea flowers.

hot and dry months. Water at the base of the plant to prevent fungal diseases on the leaves and avoid overwatering, which can lead to root rot.

Fertilizer

Yardlong beans are legumes that do not require extensive fertilization because they possess the ability to fix their own nitrogen. It is important to avoid excessive nitrogen, as this can promote leafy growth at the expense of bean pods. If the plants look pale or have stunted growth, though, a low-nitrogen fertilizer can be applied to stimulate growth.

Support

Yardlong beans are climbing vines that require support. Install a trellis, stake, pole, or tripod structure for the vine to climb.

Companion Plants

Yardlong beans grow well with corn, squash, marigolds, nasturtiums, carrots, cucumbers, radishes, brassicas, beets, lettuce, and strawberries. Avoid growing yardlong beans with onions, garlic, chives, leeks, scallions, and shallots.

Pests and Diseases

The most common pests and diseases of yardlong beans are cutworms, beetles, aphids, root rot, leaf spots, and powdery mildew. (See page 150.)

Harvest and Storage

When to Harvest

Yardlong beans typically require approximately 70 to 90 days of cultivation after sowing before they are ready to harvest. The pods will begin forming 2 to 3 weeks after flowering. It is advisable to harvest the beans when they reach a length of approximately 12 to 18 inches (30–46 cm), have a tender texture, and before the seeds begin to develop in the pods.

To harvest yardlong beans, use sharp scissors or gentle handpicking techniques to minimize damage to the plant. Regular harvesting is encouraged to stimulate future bean production.

How to Store

Yardlong beans are versatile and can be used as an ingredient in a diverse range of culinary creations, including stir-fries, soups, curries, and as a steamed side dish. Yardlong beans can be stored in the refrigerator for up to 5 days. You also can keep them in the freezer. Just blanch the beans in boiling water for 1 minute first. Then, immediately transfer them to an ice bath to stop the cooking process. Drain the beans, put them in a freezer bag, and store them in the freezer for up to 1 year.

My Favorite Vibrant Varieties

- **'MOSAIC' BEAN:** This beautiful yardlong bean has purple and red pods that are sweet, crisp, and tender. It matures in 70 to 90 days.
- **'RED NOODLE' BEAN:** This delicious yardlong bean has deep red pods that keep most of their color when cooked. The beans are ready to harvest 80 days from seeding.

↑ 'Thai Soldier' long beans are a foot (30 cm) long and very flavorful.

- **'THAI SOLDIER' BEAN:** This long bean produces firm and flavorful foot-long (30 cm) pods with dark red-brown tiger stripes. Beans mature 70 days after seeding.

'HOPI TURQUOISE' (HOPI BLUE) CORN

'Hopi Turquoise' corn is a distinct variety of blue corn, and like the majority of blue corn varieties, it is open-pollinated. 'Hopi Turquoise' corn is an ancient heirloom variety of corn traditionally grown by the Hopi people of the Southwestern United States, particularly in present-day Arizona. It is known for its striking blue to purple kernels.

Planting

Blue corn requires a sunny spot with at least 6 to 8 hours of sunlight daily. It thrives in well-drained, fertile soil with a pH between 6.0 and 6.8.

Direct-Sowing
Direct-sow the seeds in the garden in late spring or early summer once the soil temperature reaches 60°F to 65°F (16°C–18°C). Sow seeds 1 to 2 inches (2.5–5 cm) deep and 6 to 12 inches (15–30 cm) apart. Space rows approximately 18 to 36 inches (46–61 cm) apart for traditional tall blue corn varieties, reducing the spacing for shorter varieties. To enhance pollination rates, plant corn in blocks of three to four rows rather than a single row.

Care

Water
Blue corn requires approximately 1 to 1½ inches (2.5–4 cm) of water per week.

Fertilizer
'Hopi Turquoise' corn is a nitrogen-demanding crop that requires a balanced fertilizer at planting and again when the plants reach 8 to 12 inches (20–30 cm) in height. To ensure proper nitrogen uptake, I apply fish fertilizer every 3 weeks or plant blue corn alongside nitrogen-fixing plants such as beans. Upon the emergence of the tassels on the corn ears, transition to a fertilizer that is enriched in phosphorus and potassium to help kernel development.

Weeds
Remove weeds early. Hand weeding is recommended to avoid damaging the roots.

Companion Plants

Blue corn grows well with pole beans, cucumbers, squash, sunflowers, marigolds, dill, nasturtiums, melons, and borage. Avoid growing blue corn with broccoli, Brussels sprouts, cabbage, cauliflower, kale, and kohlrabi.

Pests and Diseases

The most common pests and diseases for blue corn are aphids, corn earworms, cutworms, and root rot. (See page 150.)

Harvest and Storage

When to Harvest
'Hopi Turquoise' corn is harvested in two stages. The first stage involves harvesting the corn when the kernels are fresh, milky, and tender. The second stage involves allowing the ears to dry completely on the stalk. This stage is ideal for grinding blue corn into flour or using it for decorative purposes.

How to Store
Store the completely dry kernels in an airtight container to prevent mold and pests.

'GLASS GEM' CORN

'Glass Gem' corn, renowned for its vibrant and jewel-like kernels, is a visually captivating addition to any garden. It belongs to the flint corn group, which are those that can serve as both a food source and a decorative element. As a flint corn, it is suitable for grinding into corn flour.

Planting

'Glass Gem' corn requires a sunny spot in the garden with at least 6 to 8 hours of sunlight daily. It thrives in well-drained, fertile soil with a pH between 6.0 and 6.8.

Direct-Sowing

Direct-sow the seeds in the garden after the last frost date in the spring once the soil temperature reaches 60°F to 65°F (16°C–18°C). Sow seeds 1 to 2 inches (2.5–5 cm) deep and 8 to 12 inches (20–30 cm) apart. Space the rows approximately 24 to 36 inches (61–91 cm) apart for adequate airflow and pollination. To optimize pollination efficiency, gently shake the corn stalks during the pollination period. Additionally, consider planting corn in blocks of three to four rows instead of in a single row.

Care

Water

Deeply and consistently water 'Glass Gem' corn, particularly during the dry months and during the germination and tasseling stages. 'Glass Gem' corn requires approximately 1 to 1½ inches (2.5–4 cm) of water per week.

Fertilizer

Failure to provide adequate nitrogen during the plant development stage can result in short stalks. To ensure enough nitrogen uptake, I apply fish fertilizer every 3 weeks or plant 'Glass Gem' corn alongside nitrogen-fixing plants such as beans. Upon the emergence of the tassels, transition to a fertilizer enriched in phosphorus and potassium to help kernel development.

Weeds

Weeds can cause substantial root disturbance during the germination and establishment phases. Gently remove weeds early. Hand weeding is recommended to avoid damaging the roots.

Companion Plants

'Glass Gem' corn grows well with pole beans, cucumbers, squash, sunflowers, marigolds, dill, nasturtiums, melons, and borage. Avoid growing with broccoli, Brussels sprouts, cabbages, cauliflowers, kale, and kohlrabi.

Pests and Diseases

The most common pests and diseases are aphids, corn earworms, cutworms, and root rot. (See page 150.)

Harvest and Storage

When to Harvest

'Glass Gem' corn is typically grown for its decoration, or used for popcorn or grinding into flour. Harvest once the kernels have dried and the husks have turned brown. The kernels have a high moisture content, so they need additional drying. Remove all ears from the husks and place them on a wire rack in a well-ventilated area. The ears are then dried for several more weeks until they reach the desired moisture content.

How to Store

Prior to storage, test some by popping it in a pot. If the popped kernels are well-done and taste good, they are ready. However, if the popped kernels are chewy, they need to continue drying. Store the completely dry kernels in an airtight container.

‘STRAWBERRY’ POPCORN

‘Strawberry’ popcorn is an attractive heirloom variety that produces visually appealing, dark red, jewel-like popcorn. ‘Strawberry’ popcorn stalks are a smaller size, making them an ideal choice for small garden spaces or ornamental plantings. Furthermore, ‘Strawberry’ popcorn can be used as a decorative element during the autumn season and transformed into a delectable popcorn dish for movie nights. ‘Strawberry’ popcorn is a wholesome grain food that provides an abundance of vitamins, minerals, and antioxidants.

Planting

Growing ‘Strawberry’ popcorn is quite similar to growing sweet corn. Both require full sun sites with nutrient-rich and well-drained soil. Direct-sowing the seeds in the garden is feasible in late spring or early summer once the soil temperature reaches 60°F to 65°F (16°C–18°C) and the average last frost date has passed. Popcorn needs a lengthy growing season, with some varieties requiring up to 120 days to mature.

Sow popcorn seeds 1 inch (2.5 cm) deep and 8 to 10 inches (20–25 cm) apart. Space rows approximately 18 to 24 inches (46–61 cm) apart. To enhance pollination rates, plant corn in blocks of three to four rows rather than in a single row. Maintain even soil moisture during the germination period. Popcorn germination requires a longer duration compared to sweet corn, so exercise patience.

Care

Water
‘Strawberry’ popcorn requires adequate watering to thrive. Popcorn has a shallow root system, and you may observe new roots developing above the soil at the base of the plant. These anchoring roots serve as a stabilizer for the stalk. Popcorn requires approximately 2 inches (5 cm) of water per week. Insufficient water can stress the plant during hot weather.

Fertilizer
‘Strawberry’ popcorn is a nitrogen-demanding crop. If the plant does not receive adequate nitrogen during its development, the stalk will be shorter, the ears will be smaller, and the kernels will not fully develop. I apply fish fertilizer to my popcorn every 3 weeks.

Weeds
Popcorn seedlings will not compete effectively with weeds.

‘Strawberry’ popcorn can be used as a decorative element during the autumn season and transformed into a delectable popcorn dish for movie nights.

Weeds can cause substantial root disturbance during the germination and establishment phases. To mitigate this, it is advisable to remove weeds early. Regular weed control is necessary to reduce competition for nutrients and water with the plant. Additionally, gentle hand weeding is recommended to avoid damaging the roots.

Companion Plants

'Strawberry' popcorn grows well with pole beans, cucumbers, squash, sunflowers, marigolds, dill, nasturtiums, melons, and borage. Avoid growing popcorn with broccoli, Brussels sprouts, cabbages, cauliflowers, kale, and kohlrabi.

Pests and Diseases

The most common pests and diseases of popcorn are aphids, corn earworms, cutworms, and root rot. (See page 150.)

Harvest and Storage

When to Harvest

In contrast to sweet corn, popcorn is ready for harvest once the kernels have dried and the husks have turned brown. Popcorn kernels have a high moisture content, necessitating additional drying. To fully dry popcorn, all ears are removed from the husks and placed on a wire rack in a well-ventilated area. The ears are then dried for several more weeks until they reach the desired moisture content.

↑ Colorful corn is such fun to grow.

How to Store

To ensure that the popcorn is ready for storage, test it by making some popcorn. If the popped kernels are well-done and taste good, they are ready. However, if the popped kernels are chewy, they need to continue drying. Store the completely dry kernels in an airtight container. Proper storage allows popcorn to retain its freshness for several years.

'RED BURGUNDY' OKRA

'Red Burgundy' okra is also known as red lady's finger. This variety is an open-pollinated, heirloom cultivar of *Abelmoschus esculentus*, the edible green okra. 'Red Burgundy' is a vibrantly colored okra that enhances the visual appeal of your garden and offers a range of health advantages, including its high antioxidant content and abundance of vitamins and minerals. Additionally, 'Red Burgundy' okra boasts striking hibiscus-like flowers, and both the seedpods and flowers are edible.

Planting

'Red Burgundy' okra is a visually appealing and prolific vegetable that flourishes in warm climates. It thrives in nutrient-rich, well-drained soil and requires full sun exposure with at least 6 to 8 hours of sunlight daily. Ideally, okra seeds should be direct-sown outdoors into the garden soil. However, if you reside in regions with a limited growing season, indoor seed starting is a viable option.

Sowing Indoors

Sow seeds indoors 3 to 4 weeks before the last frost date. Fill the seedling tray with a seed-starting mix and sow 1 seed per cell, ensuring the seeds are at a depth of approximately 1 inch (2.5 cm). Maintain adequate soil moisture during the germination phase. Additionally, position the seedling tray in a location with an air temperature from 65°F to 75°F (18°C–24°C). If necessary, use a heat mat to help seed germination. Once the seeds have germinated, remove the heat mat and provide them with bright, indirect sunlight to prevent leggy seedlings. The seedlings are ready for transplanting when they have developed a couple of sets of their true leaves. Before planting the seedlings outdoors, acclimate the seedlings slowly to the outdoor conditions over a period of approximately 1 week.

Direct-Sowing

Sow okra seeds directly into the garden a few weeks before the last frost date in the spring, when the soil temperature has reached approximately 65°F (18°C). Before planting, soak the seeds for approximately 12 to 24 hours to help the germination process. Sow each seed 1 inch (2.5 cm) deep and 12 to 18 inches (30–46 cm) apart. Space the rows about 3 feet (91 cm) apart. Maintain even soil moisture during the germination phase.

Care

Water

'Red Burgundy' okra requires consistent watering, particularly during flowering and pod development. They need approximately 1 inch (2.5 cm) of water per week. Deep watering once a week is recommended to foster a robust root system. This practice should be continued during the dry months. Water at the base of the plant to prevent fungal diseases and avoid over-

↑ Red okra seedlings ready for transplant

watering, which can lead to root rot. Applying a layer of organic mulch after planting will help in retaining soil moisture and suppressing weeds.

Fertilizer

'Red Burgundy' okra is not a heavy feeder. It is recommended to apply organic compost at planting time and apply a well-balanced fertilizer during the growing season.

Pruning and Support

'Red Burgundy' okra plants can grow tall, requiring the use of stakes, trellises, or cages to prevent them from toppling over. Regular pruning involves removing the lower leaves and branches to enhance air circulation and mitigate the risk of fungal diseases.

Pests and Diseases

The most common pests and diseases of okra are aphids, spider mites, caterpillars, flea beetles, stink bugs, and powdery mildew. (See page 150.)

Companion Plants

Okra grows well with tomatoes, peppers, beans, corn, cucumbers, radishes, sweet potatoes, marigolds, basil, and carrots. Avoid growing okra with potatoes, peas, snap beans, sage, and cotton.

Harvest and Storage

When to Harvest

Typically, okra is ready for harvest within 50 to 60 days after planting, although this timeframe can vary depending on the specific variety. Harvest the pods when they are young and tender, at approximately 4 to 5 inches (10–13 cm) in length. Use a sharp knife or scissors to cut the pods from the plant. Harvest the pods every 2 to 3 days to encourage future production.

How to Store

Proper storage of okra is essential for maintaining its freshness and quality.

For short-term storage, store fresh okra in a paper bag or loosely wrap it in a paper towel. Then, place it in a perforated plastic bag or container to help airflow. Store it in the refrigerator for up to 5 days.

For long-term storage, thoroughly wash and dry the okra. Trim the stems, leaving the caps intact. Blanch the okra in boiling water for a few minutes and immediately transfer it to an ice bath to halt the cooking process. Spread the okra in a single layer on a baking sheet to freeze. Once frozen, transfer them to a freezer bag. Frozen okra can last up to 12 months.

My Favorite Vibrant Varieties

- **'ALABAMA RED' OKRA:** An heirloom variety from Alabama, this okra has fat red-tinged pods that are great fried. Harvest the pods 55 days from seeding.
- **'BOWLING RED' OKRA:** 'Bowling Red' plants grow up to 8 feet (244 cm) tall. Deep red pods and stems. Great-tasting pods mature in 55 days.
- **'BURMESE' OKRA:** This heirloom variety from Burma has large pods that are tender and spineless. Pods mature in 53 days.
- **'JING ORANGE' OKRA:** This deep reddish-orange okra has some of the most tender 6-to-8-inch-long (15–20 cm) pods, and matures in 60 days.
- **'RED BURGUNDY' OKRA:** This okra has deep red pods that are very delicious and tender. It matures in 55 days.

'Red Burgundy' okra is a visually appealing and prolific vegetable that flourishes in warm climates.

2

BUILDING YOUR VEGETABLE GARDEN

Starting a garden can give you a profound sense of satisfaction, enjoyment, and happiness. Gardening is not merely a recreational activity but also a rewarding project that fosters a connection between individuals and nature. It serves as an excellent means for educating children about the origins of their food and promoting healthy eating habits.

A vibrant vegetable garden, a colorful flower garden, or even a personal oasis can all be within tangible reach! In this part of the book, we will explore building a vegetable garden, from garden placement and soil preparation to seed starting and so much more.

Consider Your Climate When You Start

Before starting on your gardening journey, it is crucial to determine your climate and hardiness zone. This knowledge is key to being able to pick the appropriate vegetables to align with your growing season, rainfall patterns, and temperature ranges.

Once you have determined your climate and hardiness zone, the next step is identifying the average first and last frost dates within your area. These dates are key to making informed decisions regarding planting, transplanting, and selecting appropriate plant varieties.

Generally, spring is the optimal season to begin gardening, provided that the soil is no longer frozen. However, certain plants may require planting in mid-spring, early or late summer, or even fall and winter.

CHOOSING THE RIGHT LOCATION

Before you start your gardening adventure, it is absolutely crucial to select an optimal location within your yard. Here are some considerations to guide you in choosing the ideal location for your garden:

Sunlight

Sunlight is a crucial component of photosynthesis and essential for plant growth. Insufficient sunlight impairs the ability of the plant to perform biological processes. However, excessive sunlight can be harmful to plants, causing stress and dehydration.

Select an area that receives at least 4 to 6 hours of sunlight daily. This will be the best site for most vegetables. Remember that different vegetables have varying sunlight requirements. For instance, broccoli needs at least 6 to 8 hours of sunlight daily and requires full sun exposure, while lettuce can tolerate that same number of hours but should be in a partial shade location.

Good Drainage

Most plants prefer well-drained soil. If the roots of most vegetables are constantly wet, they can develop root rot. If you have a suitable garden space but it has poor drainage, raised beds can be the answer to improved drainage.

Nutrient-Rich Soil

Healthy plants thrive in well-nourished soil. If your soil lacks essential nutrients, consider adding plenty of organic matter such as compost to enhance the soil structure and help plant growth.

↑ The south-side garden in spring

WHAT TYPE OF GARDEN SHOULD I MAKE?

Whether you reside in a spacious home with a large yard, live in a townhouse with a nearby community garden plot, or rent an apartment with limited window and balcony space, there are many types of gardens available for you to explore for your new vegetable garden.

Raised Bed Gardens

What Is a Raised Bed?
Raised beds are freestanding garden beds a few feet tall constructed on top of the ground, resulting in a soil level that is higher than the surrounding area. These beds can be constructed out of many different materials such as rot-resistant lumber, bricks, concrete blocks, or metal.

Raised garden beds offer many benefits:

- **Enhanced Soil Conditions:** A raised bed has soil that is less compacted, promoting freer movement of water and air through the soil.
- **Less Maintenance:** Weed control is simple due to the height of the bed (usually within reach). Often vegetable crops can be planted closer together in raised beds, which creates shade and reduces moisture loss. Pest control becomes more manageable

MY SECRET FORMULA FOR SOIL AMENDMENTS IN RAISED BEDS

- **Step 1.** After the frame has been constructed and before the soil is added, lay weed-resistant fabric on the bottom of your raised bed, followed by a layer of cardboard.
- **Step 2.** Collect enough wood, twigs, branches, vegetable stalks, and kitchen scraps to half-fill the raised beds. This practice will help you save money by reducing the amount of soil you will need.
- **Step 3.** Incorporate a good quality, organic, raised bed soil or quality topsoil, then add perlite and organic compost into the soil mixture.
- **Step 4.** Top the soil with worm castings, an all-purpose organic fertilizer, and a rock dust or soil remineralizer.
- **Step 5.** Add water to the raised bed as required for seed germination or plant growth. Adequate water is essential for good seed germination and growth of your favorite vegetables.

↑ There are many different sizes of containers to grow your vibrant vegetables.

because plants are easier to inspect and pests can be handpicked. Additionally, wire netting can be placed at the bottom of the raised bed to prevent rodents from tunneling into the bed. The raised bed frame can also serve as a support for row covers that can be added for frost or shade protection.

Can Raised Beds Be Placed Directly on Concrete?
I have successfully placed several raised beds directly on concrete. And the vegetables growing in these raised beds are thriving, outperforming those planted directly in the ground nearby.

↑ Setting up a raised bed on concrete is a great way to grow plants in urban areas or places with limited soil access.

↑ Container gardening is an efficient way to grow plants in limited spaces.

Container Gardening

A container garden is an excellent choice for small outdoor spaces, particularly balconies, as the containers can fit the space available. No need to haul in a big load of soil. Just bring in the number of bags needed to fill the pots. Container gardening isn't limiting and allows for the cultivation of a variety of vegetables, including tomatoes, eggplants, peppers, lettuce, and herbs, among others. To ensure optimal growth, place the containers in a location that receives full sunlight. Additionally, elevate the containers above the ground to deter pests and enhance drainage.

Choose the Right Container

Containers offer versatility in gardening, enabling the cultivation of a range of plants. The selection of appropriate container sizes is crucial for optimal growth. Larger containers provide ample soil space, better soil moisture retention, and more room for extensive root development.

Vegetables generally require containers that are at least 5 gallons (19 L) in size, while herbs typically need containers ranging from 1 to 3 gallons (4–11 L). Tomatoes, cucumbers, pole beans, snap peas, and squash plants will need support, such as trellises, and should have deeper containers for better results. Additionally, root vegetables will benefit from containers with sufficient depth.

Do-It-Yourself (DIY) Containers

When I first started my gardening journey, I adopted a DIY approach to container creation. I procured storage containers or plastic buckets from hardware stores and meticulously drilled numerous drainage holes in the bottom.

It is crucial to underscore the importance of drainage in maintaining the health of vegetable plants. Wet feet and soggy soil can easily lead to root rot and plant demise.

PLASTIC POTS	

Pros	**Cons**
Keep soil moist longer	Prone to tipping over
Many colors available	At the end of their lifespan, they create plastic waste
Strong and durable	Plastic can become brittle and crack over time when exposed to cold and UV light
Lightweight	
Inexpensive	
Most plastic containers are made from recycled materials	
Easy to add extra drainage holes	

TERRACOTTA CLAY POTS	

Pros	**Cons**
Durable natural material	Don't hold moisture well
Look more traditional and aesthetically pleasing	Easily breakable
Heavy, but more stable	Heavy, which makes them hard to move around
Insulate and protect plant roots from stress and damage due to temperature changes	Harder to add extra drainage holes
	More expensive

↑ Vertical gardening is a great method for maximizing and optimizing the functionality of a small-scale garden.

Vertical Gardens

A vertical garden is a method of growing plants vertically, typically in stacked towers, hanging grow bags on fences or walls, or trellises or arches. Vertical gardening is a great solution for maximizing space, particularly in small outdoor areas. It is well-suited for balconies because you can grow vegetables vertically, which optimizes space utilization. Additionally, it requires less soil compared to traditional gardening methods.

Stacked Tower Gardens

Stacked towers are constructed from durable plastic and are an efficient space-saving solution for small gardens, balconies, patios, and greenhouses. They enable gardeners to garden vertically, allowing for the cultivation of a range of plants, including salad greens, Swiss chard, strawberries, collard greens, herbs, and numerous varieties of more compact vegetables.

Wall Gardens

Pocket vertical grow bags are an excellent choice for wall gardens due to their simplicity and cost-effectiveness. However, it is important to note that each pocket can only hold a limited amount of soil, which may result in faster drying of plant root balls. Therefore, it is crucial to water the plants approximately 4 to 5 times per week, or even daily, depending on the specific vegetable variety. Wall grow bags are particularly suited for growing salad greens, herbs, strawberries, and smaller flowers.

Trellis and Arch Gardens

Trellises and arches are widely recognized vertical elements in gardens. They offer ample support for plants and can be conveniently attached to the ground, containers, or raised beds. Numerous affordable trellises and arches are readily available to purchase online. Pole beans, long beans, vining squash, cucumbers, indeterminate (vining) tomatoes, malabar spinach, bitter melons, and snap peas are among the most suitable edible plants that benefit from growing on trellises.

In-Ground Gardens

In-ground gardening has been a practice for hundreds of years, offering a cost-effective alternative to other gardening methods. While it may require more initial investment, in-ground gardens can be a rewarding and low-maintenance option for those seeking to transform a portion of their lawn into a vibrant garden space.

- **Step 1. Select a suitable location.** Choose a raised area in your yard that remains dry during the rainy season.
- **Step 2. Prepare the area.** Remove the existing grass and vegetation from the selected area. For effective grass removal, lay clear plastic over the grass for several weeks during hot weather. Then, manually remove the dead grass.
- **Step 3. Prepare the soil.** Use a shovel to loosen the soil to a depth of at least 6 to 10 inches (15–25 cm) to reduce soil compaction. Test the soil to determine its nutrient and pH levels. Apply organic matter such as compost and kitchen scraps to enhance the soil fertility.
- **Step 4. Plant your vegetables.** With the soil prepared, you are now ready to plant your vegetable garden.

↑ Corn plants growing in an in-ground garden

In-ground gardens can be a rewarding and low-maintenance option for those seeking to transform a portion of their lawn into a vibrant garden space.

PREPARING THE SOIL

Healthy soil is the foundation of a successful garden. Well-amended soil provides essential nutrients to the plants, supporting their growth and development. Here are some ways to improve soil health:

Know Your Soil Type and Texture

If you intend to plant in the ground, it is essential to assess your soil texture. A simple test involves taking a clump of soil, slightly wetting it in your hand, and determining its consistency. Excessively sandy or clay soil can be amended for better plant growth.

- **Sandy soil** will be gritty due to the presence of small rocks and will be prone to crumbling.
- **Clay soil** will be sticky when wet and form a ball when squeezed.

Test Soil pH

A pH test will determine the acidity or alkalinity of your soil. Most plants thrive in a neutral pH range of 5 to 6. If you suspect a pH imbalance (too acidic or too alkaline), you can acquire a pH test kit from your local nursery or send a soil sample to a soil testing lab.

Increase Soil Aeration

Soil requires oxygen for optimum plant growth. Well-aerated soils will have a better soil texture. To increase soil aeration, incorporate organic matter and prevent soil compaction by using raised beds or containers.

Add Organic Matter and Improve Soil Texture

Incorporating 2 to 4 inches (5–10 cm) of organic compost is an effective method of enhancing your soil structure. This can be easily done in raised beds or containers.

Mulch the Soil

Applying a layer of mulch over the topsoil is an effective method to enhance water retention and suppress weed growth. Organic mulch options include garden straw, leaves, bark mulch, wood chips, and grass clippings. Keep in mind that the grass clippings must not be treated with pesticides or chemical fertilizers.

↑ Healthy soil, ready for planting

CHOOSING THE RIGHT PLANTS

Selecting the appropriate plant species for your location requires understanding their specific growing requirements. Plants are categorized into three main groups based on their life cycle.

- **Annuals:** These plants undergo their entire life cycle within a single growing season and are easily killed by frost. They need to be started again from seed each year. Examples of annuals include beans, broccoli, cabbage, cucumbers, eggplants, lettuce, peas, peppers, radishes, spinach, tomatoes, and zucchini.
- **Perennials:** Perennials may retain their foliage through the winter or they may lose them and go dormant, but the plant will regrow from its roots each year. Examples include horseradish, lovage, asparagus, and rhubarb.
- **Biennials:** Biennials exhibit a unique life cycle. They produce foliage during their first year and flower or bear fruit during their second year, if grown in a suitable climate. After fruiting, they die. Examples include beets, Brussels sprouts, carrots, kale, kohlrabi, leeks, onions, and parsley.

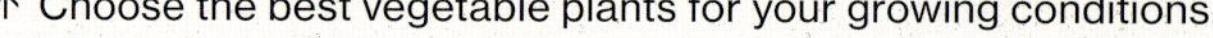

↑ Choose the best vegetable plants for your growing conditions.

Crops for Beginners

Cultivating vegetables is a rewarding and enjoyable endeavor. To enhance your chances of success, it is advisable to begin with a manageable number of crops and gain expertise in their cultivation.

For beginners, a recommended approach is to focus on a limited selection of easy-to-grow vegetables and flowers.

Here are some of the most suitable plants for beginners:

- Beans
- Beets
- Carrots
- Cilantro
- Lettuce
- Peppers
- Radishes
- Snap peas
- Spinach
- Swiss chard
- Tomatoes
- Zucchini
- Calendulas
- Cosmos
- Marigolds
- Zinnias

Grow What You Eat

The most effective strategy for success is to cultivate crops that align with your family's eating preferences. If your family enjoys a particular vegetable, such as tomato, then plant an abundance of it and put an additional effort in to maximize its yield. Additionally, you can preserve a portion of your summer tomato harvest to provide sustenance throughout the year. Conversely, if a vegetable like okra is not favored by most of your family, don't waste your time and space cultivating it in your first vegetable garden.

CHOOSING A COLOR PALETTE

When designing my garden, I always consider the impact of colors. The colors of a garden can reflect your energy and immediately influence your mood. I find it aesthetically pleasing to combine various colors of flowers with different colors of vegetables.

The more colors—the better!

Determine Your Desired Ambiance

The initial step is to determine the desired ambiance that you want for your garden. Do you envision a tranquil and soothing space or a vibrant and energizing one? I employ color schemes that align with the changing seasons. For instance, in spring, I cultivate a variety of pink, yellow, white, purple, and pastel flowers that complement the green vegetables. In summer, I opt for bright red, orange, and hot pink colors. And in the autumn, I incorporate dark pink, burnt orange, white, green, burgundy, and dark red colors.

Pick a Color Scheme

If you are uncertain about picking your color palette, you can use the color wheel. The color wheel is composed of twelve colors. There are three primary colors: yellow, red, and blue. There are three secondary colors: purple, green, and orange. Last, there are six tertiary colors. If you envision a soothing and calming garden, opt for colors with less contrast, such as green, light pink, purple, and yellow. Conversely, if you desire a vibrant and energizing garden, select high-contrast colors, including red, blue, and green, along with some bright pink, orange, and yellow.

Color Scheme Examples

An **analogous** color scheme has a combination of three colors that are found next to each other on the color wheel. For example: purple, blueish purple, and blue.

A **monochromatic** color scheme has a base of just one color—so you don't need to use the color wheel. Just pick one color and run with it. For example: yellow bell peppers with bright yellow Swiss chard and yellow cherry tomatoes.

A **complementary** color scheme consists of two colors that are opposite each other on the color wheel. For example: blue and orange, or purple and yellow.

A **complex** color scheme basically combines all of the above together. I personally like the complex color schemes of summer—the vegetable garden brings so much joy, vibrant feelings, and energizing energy during this season.

→ There are twelve colors on the color wheel.

↑ White flowers provide a refreshing contrast on hot days.

Adding White

Why consider the color white? Despite its often overlooked and perceived colorlessness, white serves as a versatile color that can complement any color scheme. White provides a refreshing contrast during hot summer days and enhances the ambiance of shady areas—adding an air of elegance to any garden. Throughout the year, I like to incorporate white flowers such as Iceberg roses, agapanthus, cosmos, and white vegetables like cauliflower into my garden, appreciating their ability to bring a touch of sophistication and tranquility to the color scheme.

STARTING YOUR PLANTS

At the outset of my gardening endeavors, I visited my local nursery, where I made the decision to purchase seedlings for a diverse range of vegetables, including peppers, tomatoes, beans, carrots, and squash. As a novice gardener, I had not previously considered the option of growing from seed. I was under the impression that it was a time-consuming process that required a lot of supplies. Once I tried it, growing plants from seeds indoors to transplant provided me with a sense of immediate gratification, as the plants flourished and grew rapidly—whereas I found direct-sowing could be a protracted process. However, sometimes transplanting needs significant space and, in certain instances, yielded unfavorable outcomes, depending on the specific plants involved.

During my second year of gardening, I gained valuable knowledge, discovering that certain plants are best suited for germination from seeds, while others thrived as transplants.

DIRECT-SOWING VERSUS GROWING FROM TRANSPLANTS

Regardless of your gardening experience, it's crucial to understand the differences between transplanting and direct-sowing. The method you choose depends on the specific crops you intend to cultivate, as well as your climate conditions—including the first and last frost dates.

Direct-Sowing

Direct-seeding and direct-sowing are two terms that describe the practice of planting seeds directly into the outdoor garden soil. Direct-sowing eliminates the need for having seed-starting equipment and a suitable indoor spot for germination and growing the seedlings. Direct-seeding closely resembles the natural germination process, while also providing gardeners the opportunity to select from a range of varieties. Also, direct-sowing is particularly advantageous for plants that are sensitive to root disturbance during transplanting.

Timing is a crucial factor. Gardeners should consider their average first and last frost dates when deciding on the specific types of plants they wish to cultivate. It is essential to consult local climate data to determine the optimal time for direct-sowing in your own garden.

While direct-sowing offers numerous benefits, it does carry a higher risk of plant mortality compared to using transplants. Seeds (and young seedlings) are exposed to external weather conditions such as flooding,

drought, and strong winds, which can potentially disrupt the growth and yields of the plants, leading to crop failures.

Benefits of direct-sowing:

- **Saves Time and Effort:** Direct-sowing eliminates the necessity of starting your plants indoors.
- **Promotes Stronger Root Systems:** Planting seeds directly into the soil allows the plant roots to grow and establish without being subject to damage or disrupting the roots when transplanted.
- **Saves Money:** Direct-sowing is cost-effective because it eliminates the expense of supplies and equipment needed for seed starting.

Here is a list of some vegetables that always perform better when direct-sown:

- Beans
- Carrots
- Parsnips
- Snap peas
- Spinach
- Turnips

Here are some vegetables that can either be direct-sown or grown from transplants:

- Beets
- Corn
- Cucumbers
- Collards
- Kale
- Lettuce
- Pumpkins
- Summer squash
- Swiss chard
- Watermelon

↑ Transplants

↑ Direct seeding

Growing from Transplants

Growing from transplants offers several advantages, including enhanced control and predictable outcomes. Using transplants also provides a head start on the outdoor growing season by enabling the plants to mature sooner, resulting in an earlier harvest.

If you choose to grow your transplants from seeds indoors, it is imperative to acclimate your seedlings by progressively exposing them to longer and longer hours of outdoor sunlight and wind exposure each day. This process, called hardening off, is particularly advantageous for gardeners residing in northern regions with limited growing seasons.

Benefits of growing from transplants:

- **Healthier Plants:** Transplants are healthier and mature sooner when compared to direct-sowing the same plant in the outdoor garden.
- **Succession-Planting:** You can succession-plant crops several times per season for a bigger and continuous harvest.
- **Extend the Growing Season:** You can start seeds earlier indoors and harvest sooner.

Here are some vegetables that are best grown from transplants:

- Broccoli
- Brussels sprouts
- Cabbage
- Cauliflower
- Eggplants
- Kohlrabi
- Leeks
- Peppers
- Tomatoes

WHEN TO START SEEDS INDOORS

Starting seeds indoors provides your plants with a competitive advantage, particularly in regions with limited growing seasons. The timing of seed sowing is dependent on the specific type of vegetable you intend to grow and the weather conditions in your location.

Below is essential information to help you in determining the optimal date to start sowing your seeds within your geographical region.

Know Your Plant Hardiness Zone

For slower-growing vegetables, such as tomatoes, peppers, and eggplants, it is recommended to initiate seed-starting indoors approximately 6 to 8 weeks before the anticipated average last frost date in the spring. Conversely, for rapidly growing crops like cucumbers, squash, and herbs, it is advisable to sow seeds indoors 4 to 6 weeks before the last frost date in the spring. Finally, for the swift-growing leafy greens like lettuce and spinach, it is recommended to start seeds indoors 2 to 4 weeks before the last frost date.

Consult Your Local Gardening Resources

For advice appropriate to your area, contact your local agricultural extension office, garden clubs, nurseries, and experienced gardening neighbors.

Read the Seed Packet

Most seed packets provide instructions on when the seeds should be started indoors or sown directly outdoors, typically in relation to the anticipated frost date.

Collecting Historical Weather Data

If you can, it is handy to have historical temperature data spanning at least 10 to 30 years for your geographical region. Your focus should be on the minimum daily temperatures recorded during the spring season (and specifically, the dates of the last frosts) and the fall season (corresponding to the date of the first frost). Once you have the historical record of your last frost dates, an average frost date can be determined. (See the sidebar at 142.)

↑ Me with a tray of seedlings ready to plant

HOW TO CALCULATE THE AVERAGE FROST DATES

- **Step 1.** Convert each historic frost date into the number of days it is from January 1. For example, April 23 would be day 112 of the year and May 7 would be day 126.
- **Step 2.** Add together the last frost dates in the spring for all the years available and divide by the number of years.
- **Step 3.** Do the same steps for the first frost dates in the fall for all the years and divide by the number of years.

Example:
For the last frost dates: (April 15 + April 25 + May 2) ÷ 3 = (Day 104 + Day 114 + Day 121) ÷ 3 = the average last frost date is Day 113 or April 23

For the first frost dates: (October 10 + October 5 + September 30) ÷ 3 = (Day 283 + Day 278 + Day 273) ÷ 3 = the average first frost date is Day 278 or October 5

What Is the Frost Date?

The frost date is the average date in your region when the temperature drops to or below 32°F (0°C). It serves as a crucial reference for planning sowing, planting, and harvesting activities in the garden. Additionally, it enables gardeners to plan for installing row covers for extra frost protection or even moving some plants indoors.

There are two types of frost dates:

- **Last Frost Date in Spring:** This is the average last day in the spring when frost is anticipated, indicating that it is generally safe to plant tender, frost-sensitive crops outdoors after this date.
- **First Frost Date in Fall:** This is the average first day of frost in the fall, which marks the end of the growing season for many plants.

WHEN TO START SEEDS OUTDOORS

Direct-sowing refers to the practice of planting seeds directly into the ground in a garden, rather than starting them indoors in pots. The optimal time for direct-sowing depends on the specific plant species, your local last frost date, and the plant's preferred temperature range. This method is particularly suitable for hardier plants that can withstand cooler temperatures and outdoor conditions. Still, seeds are best sown when the soil temperature is sufficiently warm for optimum germination.

Use the Last Frost Date

For cool-season crops, sow the seeds outdoors 2 to 4 weeks before the anticipated last frost date. These are crops that can withstand a light frost, such as lettuce, spinach, kale, radishes, carrots, peas, and broccoli.

For warm-season crops, sow the seeds outdoors after the average last frost date has passed and once the soil has warmed sufficiently. Examples include beans, cucumbers, squash, melons, and corn.

Seed Germination and Soil Temperature

Certain seeds require specific soil temperatures for germination. Cool-season crops germinate optimally at temperatures between 45°F and 65°F (7°C–18°C), while warm-season crops require soil temperatures ranging from 65°F to 85°F (18°C–29°C). Before planting, it is advisable to use a soil thermometer to determine when you have reached the appropriate soil temperature.

Weather Considerations

Wait for a period of consistently warm weather and avoid planting just before a cold snap. Ensure that the soil is dry enough to work so you don't have compaction and rotting issues. Overly wet soil can hinder seed germination significantly.

↑ Read each seed packet prior to planting.

UNDERSTANDING THE SEED PACKET

Knowing how to read seed packets is essential for successful gardening, as they offer crucial information on planting, cultivation, and caring for your plants. Each seed company designs and lays out its seed packaging differently, but generally, each seed packet should contain a description of the seed being sold, the recommended planting time, instructions on how to plant the seeds, and what to expect from the resulting plants. Here is how to interpret a seed packet.

The Front of the Seed Packet

The front of a seed packet has the following information:

- **Plant Name:** The packet lists the plant name and may include both the common name and scientific botanical name.
- **Picture:** The packet will often have a plant image or drawing showing the mature plant or fruit for identification purposes.
- **Type of Plant:** The packet will identify whether the plant is an heirloom, hybrid, organic, and so on.

The Back of the Seed Packet

The back of a seed packet has the following information:

- **Plant Description:** The seed packet will have a comprehensive description of the life cycle, including its classification as an annual, perennial, or biennial. Additionally, the packet will include information about the historical background of the plant, mature height, and width.
- **Days to Germination:** To help with scheduling, the packet will show the number of days it takes for the seeds to sprout under optimal conditions.
- **Planting Depth:** There will be a recommendation on how deep to plant the seeds, usually listed in inches or centimeters.
- **Planting Space:** Save the seed packet after sowing, especially if seeding indoors. You will need to know the distance between plants and rows to ensure proper spacing when transplanting into the garden.
- **Light Requirements:** There will be a handy reminder about the amount of sunlight required by the plant, usually listed as full sun, partial sun/shade, or shade.
- **Days to Maturity:** You'll find information about the time from sowing or planting to harvest. This is highly useful for planning your gardening season.
- **Plant Size:** The seed packet will also give information about the expected mature height of the plant, which helps with garden layout planning.
- **Hardiness Zone:** The hardiness rating for the plant helps gardeners to determine the plants that are most likely to thrive in their location.
- **Seed Starting:** Often the back of the seed packet will indicate whether seeds should be initiated indoors or direct-sown outdoors, typically in relation to the anticipated last frost date.
- **Special Notes:** Depending on the seed company, some packets may include harvesting tips, pollinator information, plant support ideas, pruning techniques, and fertilizer recommendations.

SEED SOWING CHART

VEGETABLE	INDOOR SOWING	DIRECT-SOWING	GERMINATION TIME	SPACING	DAYS TO HARVEST
Artichoke	8–10 weeks before last frost	After the last frost	10–14 days	24–36 inches (61–91 cm)	150–180 days
Arugula	Not recommended	Early spring/fall	5–7 days	4–6 inches (10–15 cm)	30–50 days
Asparagus	Not recommended	Early spring	10–14 days	12–18 inches (30–46 cm)	2–3 years
Basil	4–6 weeks before last frost	After the last frost	5–10 days	8–12 inches (20–30 cm)	50–60 days
Bean (Bush)	Not recommended	After the last frost	7–10 days	4–6 inches (10–15 cm)	50–60 days
Bean (Pole)	Not recommended	After the last frost	7–10 days	6–12 inches (15–30 cm)	60–70 days
Beet	Not recommended	Early spring/fall	5–10 days	3–4 inches (7.5–10 cm)	50–70 days
Bok Choy	4–6 weeks before last frost	Early spring/fall	7–10 days	6–8 inches (15–20 cm)	45–60 days
Broccoli	6–8 weeks before last frost	After the last frost	7–10 days	18–24 inches (46–61 cm)	70–100 days
Brussels Sprout	6–8 weeks before last frost	Early spring	7–10 days	18–24 inches (46–61 cm)	90–120 days
Cabbage	6–8 weeks before last frost	Early spring	7–12 days	18–24 inches (46–61 cm)	60–100 days
Cabbage (Chinese)	4–6 weeks before last frost	Early spring/fall	5–10 days	12 inches (30 cm)	60–80 days
Carrot	Not recommended	Early spring/fall	10–20 days	2–3 inches (5–7.5 cm)	70–80 days
Cauliflower	6–8 weeks before last frost	After the last frost	7–10 days	18–24 inches (46–61 cm)	60–80 days
Celery	10–12 weeks before last frost	Not recommended	14–21 days	6–8 inches (15–20 cm)	120–140 days
Corn (Sweet)	Not recommended	After the last frost	7–10 days	12–15 inches (30–38 cm)	60–100 days
Cucumber	3–4 weeks before last frost	After the last frost	7–10 days	12–24 inches (30–61 cm)	50–70 days
Eggplant	8–10 weeks before last frost	After the last frost	7–14 days	18–24 inches (46–61 cm)	70–90 days
Fennel (Bulb)	6–8 weeks before last frost	After the last frost	7–10 days	12 inches (30 cm)	90–115 days
Garlic	Not recommended	Fall (cloves)	7–14 days	6 inches (15 cm)	120–150 days

VEGETABLE	INDOOR SOWING	DIRECT-SOWING	GERMINATION TIME	SPACING	DAYS TO HARVEST
Greens (Collard)	6–8 weeks before last frost	Early spring/fall	7–14 days	18 inches (46 cm)	60–80 days
Greens (Mustard)	4–6 weeks before last frost	Early spring/fall	5–10 days	6–12 inches (15–30 cm)	40–50 days
Kale	4–6 weeks before last frost	Early spring/fall	5–10 days	12–18 inches (30–46 cm)	50–75 days
Kohlrabi	6–8 weeks before last frost	Early spring/fall	7–10 days	8–12 inches (20–30 cm)	50–70 days
Lettuce	4 weeks before last frost	Early spring/fall	7–10 days	6–12 inches (15–30 cm)	40–60 days
Okra	4–6 weeks before last frost	After the last frost	7–14 days	12–18 inches (30–46 cm)	50–65 days
Onion (Bulb)	8–10 weeks before last frost	Early spring (sets/seeds)	7–10 days	4–6 inches (10–15 cm)	90–120 days
Parsley	6–8 weeks before last frost	After the last frost	14–21 days	8–10 inches (20–25 cm)	70–90 days
Pea (Snap)	Not recommended	Early spring	7–14 days	2–3 inches (5–7.5 cm)	60–70 days
Pepper (Bell)	8–10 weeks before last frost	After the last frost	7–14 days	18–24 inches (46–61 cm)	60–90 days
Potato (Sweet)	Not recommended	After the last frost (slips)	7–14 days	12–18 inches (30–46 cm)	90–120 days
Pumpkin	Not recommended	After the last frost	7–14 days	36–48 inches (91–122 cm)	90–120 days
Radish	Not recommended	Early spring/fall	4–7 days	2 inches (5 cm)	20–30 days
Scallion	8–10 weeks before last frost	Early spring	7–14 days	2–3 inches (5–7.5 cm)	60–80 days
Spinach	Not recommended	Early spring/fall	7–14 days	6 inches (15 cm)	40–50 days
Spinach (Malabar)	4–6 weeks before last frost	After the last frost	10–20 days	12 inches (30 cm)	70–80 days
Swiss Chard	Not recommended	Early spring/fall	7–14 days	12 inches (30 cm)	50–60 days
Tomato	6–8 weeks before last frost	After the last frost	5–10 days	24–36 inches (61–91 cm)	60–80 days
Watermelon	4–6 weeks before last frost	After the last frost	5–10 days	36–48 inches (91–122 cm)	80–100 days
Zucchini	Not recommended	After the last frost	5–10 days	24–36 inches (61–91 cm)	50–60 days

↑ Seed-starting tools include a heat mat, seed starter tray with humidity dome, LED grow light, seeding tools, plant labels, and seeds.

DIY SEED-STARTING MIX RECIPE

- **1 part peat moss or coconut (coco) coir.** Seed-starting mixes that contain peat moss or coconut coir have better moisture retention and a lighter soil structure—making it easy for delicate seedling roots to penetrate.
- **1 part perlite or coarse sand.** Perlite or coarse sand helps improve aeration and drainage in the seed-starting mix.
- **1 part vermiculite.** Vermiculite helps retain moisture and nutrients in the seed-starting mix.

Mix parts thoroughly and moisten before using.

SEED-STARTING TOOLS

Starting seeds requires a few basic tools and supplies to ensure healthy seedings and successful germination.

Essential Seed Starting Tools

Seed Trays or Containers

Specialized seed trays, plug trays, or any clean container with drainage holes, such as yogurt cups or egg cartons, can be used for this purpose.

Seed-Starting Mix

Seed-starting mix is a soilless growing medium specifically formulated to support seed germination and early seedling growth. It is lightweight, sterile, and well-draining to provide the ideal conditions for seeds to sprout and develop strong roots.

To use a seed-starting mix:

- **Step 1.** Lightly moisten the seed-starting mix with warm water.
- **Step 2.** Fill the seed trays or containers with the mix.
- **Step 3.** Plant seeds according to the seed packet depth instructions.
- **Step 4.** Water gently and cover with a humidity dome or clear plastic bag if more humidity is needed.

Labels or Markers

Plant labels are essential for keeping track of what you've planted and where. They will help you identify plants during their early growth stages. This is especially helpful with many brassica seedlings that can look alike. Labels will help keep your crops organized.

Spray Bottle
A plant spray bottle is a handy tool for watering, misting, and applying fertilizer treatments to your plants. It's especially useful for watering delicate seedlings and maintaining humidity levels for overwintering tropical vegetable plants.

Grow Lights
Grow lights are artificial lights designed to support plant growth indoors by mimicking sunlight. They provide the necessary spectrum of light for photosynthesis, which is crucial for seedlings, houseplants, and indoor gardens.

Heat Mat
A heat mat is a device that provides consistent, gentle heat to the bottom of the seed-starting trays or pots to encourage even seed germination and healthy root growth. It's especially useful for plants that require warm soil to thrive, like hot peppers.

Humidity Dome
A humidity dome is a clear plastic cover placed over plants, seedlings, or seed trays to maintain a high level of humidity and create a greenhouse-like environment. This is especially useful for promoting seed germination and protecting young plants from drying out.

Seeds
Seeds are the starting point for growing plants, whether you're cultivating vegetables, flowers, herbs, or trees. Starting from seeds will allow you to grow a variety of plants, often at a lower cost.

Optional but Useful Tools

Seedling Fertilizer
This is a specially formulated fertilizer to provide the nutrients young plants need for strong, healthy growth during their early seedling stage.

Soil Thermometer
A soil thermometer is a tool used to measure the temperature of the soil, which is an essential factor in gardening. Soil temperature can affect seed germination, root growth, and overall plant health.

Transplanting Tools
These gardening tools are designed to help safely and efficiently move young plants or seedlings from one location to another, such as from seed trays to larger containers or to the garden. This tool helps minimize root disturbance and improve the chances of successful transplantation.

Fan or Ventilation System
Fans and vents are used to enhance air circulation around plants, usually in controlled environments such as greenhouses, grow tents, polytunnels, or indoor gardens. Proper ventilation is crucial for promoting healthy plant growth by regulating temperatures, humidity, and carbon dioxide levels. Fans and vents also prevent the accumulation of heat and excessive moisture, which can lead to plant stress and fungal diseases.

Self-Watering Seed Trays
Self-watering seed trays are designed to help maintain consistent moisture levels for seedlings by automatically supplying water to their roots. These trays are commonly used for seedlings, young plants, or houseplants that are being grown in containers or pots. A self-watering tray can also help reduce the frequency of watering, prevent overwatering, and ensure that plants receive consistent hydration, which is especially useful when you are away or need extra help during the hot weather.

Light Timer
A timer for grow lights is a useful tool that will automatically control the on/off cycle of your grow lights, ensuring that your plants receive the right amount of light each day. Consistent light exposure is crucial for plant health, particularly at the seedling stage.

DIY Alternatives

- Repurposed containers, such as egg cartons, paper cups, or milk cartons, can be used for seed starting.
- Use plastic wrap or plastic bags to cover the containers to maintain humidity if you don't have a humidity dome.

3

GARDEN CARE

Proper garden care is essential for maintaining the health, beauty, and productivity of outdoor spaces. A well-maintained garden not only enhances the aesthetic appeal of a home but also contributes to environmental sustainability and personal well-being. Garden care involves several interconnected practices, including pest management, weed control, fertilizing, watering, and general plant care—all working together to ensure the garden remains vibrant and sustainable.

In this section, you'll discover that garden care is more than just a chore; it's a rewarding practice that brings beauty, health, and sustainability to your surroundings. By embracing thoughtful planning, regular maintenance, and eco-friendly methods, anyone can create and maintain a flourishing garden that will benefit people and the planet.

MANAGING PESTS AND DISEASES

It is important to scout for harmful insects and diseases that can damage crops and flowers. Integrated pest management (IPM) is a widely used approach that combines preventive techniques like companion planting and crop rotation with natural solutions such as introducing beneficial insects like ladybugs. Organic methods, such as neem oil sprays or homemade garlic solutions, can also be effective while being environmentally friendly.

VEGETABLE GARDEN PESTS

PEST	SIGNS/SYMPTOMS	CONTROL/PREVENTION
Aphids: Tiny, soft-bodied, oval-shaped, sap-sucking insects. They are a member of the insect superfamily Aphidoidea.	Aphids cause deformation of leaves, flowers, and buds, and leave a sticky glaze like honey on the leaves. You will notice a lot of ants on the plant too.	• Spray the aphids with a jet of water. • Use an insecticidal soap spray. Mix a few drops in 1 gallon (4 L) of water. Spray 2–4 times a week for a couple of weeks. • Grow companion plants such as nasturtiums. • Release beneficial insects such as ladybugs.
Cabbage loopers and cabbage worms: Small, smooth, white-striped and green caterpillars are commonly found on cabbages, collards, broccoli, cauliflower, and kale.	These caterpillars are one of the most destructive pests in the garden. They typically eat voraciously and make many holes in the leaves. They can also cause stunted growth of the plant.	• Handpick the caterpillars. • Use Bt (*Bacillus thuringiensis*) spray: Mix 1 tablespoon (15 ml) Bt with 1 gallon (4 L) of water and spray on the top and bottom of the leaves.
Corn earworms: Also known as fruitworms, the larvae grow about 1–2 inches (2.5–5 cm) long and are green to light brown to dark brown in color. In addition to corn, they also attack many other vegetable and fruit plants such as lettuce, broccoli, cantaloupe, and tomatoes.	Earworms feed on corn foliage, silk, and kernels, leaving behind a trail of destruction.	• Use Bt (*Bacillus thuringiensis*) spray. • Handpick the caterpillars. • Apply mineral oil on the corn silk. • Release beneficial insects such as ladybugs.
Cutworms: These caterpillars are brownish black with a stripe across their body. They usually curl their body into a "C" when they are disturbed.	You'll observe wilted plants, severe damage to stems, and sometimes young plants that are cut off completely near the ground level.	• Handpick the caterpillars. • Rototill the soil to destroy larvae. • Use cardboard collars around young plants. Push the collar a couple of inches into the soil to create a barrier. • Clear out weeds and plant residue around your vegetable crops.
Earwigs: This insect is also known as a pincher bug because it has two long claw-like appendages at the end of its body.	Earwigs feed on plant leaves, decaying wood, and rotting vegetables, flowers, and fruits.	• Remove plant debris. • Spray rubbing alcohol directly on the earwig (not on the plants). • Eliminate excessive moisture and keep the garden clean.

PEST		SIGNS/SYMPTOMS	CONTROL/PREVENTION
	Japanese beetles (adults and grubs): Adult Japanese beetles have metallic green heads and shiny brown wings. The larvae or grub stage has a soft, cream-colored body.	Adult Japanese beetles feed on the leaves, flowers, and fruits of many plants. On the other hand, the grubs feed on roots, and can often be found under the grass, and in raised beds or containers.	• Use a neem oil spray. • Handpick the adult beetles and grubs. • Use milky spore disease sprays and beneficial nematodes to control the larvae. • Cover your plants with row covers. • Grow trap plants to attract the beetles away from your crops.
	Leaf miners: These are small, white-yellow maggots or larvae.	Leaf miners feed on leaves and create tunnels beneath the foliage surface.	• Remove the affected leaves. • Rototill your soil to destroy the pupae. • Remove plant debris.
	Snails and slugs: Slugs are small, soft-bodied, slimy mollusks. Snails are similar and have a shell. They mostly like moist conditions and come out at nighttime.	Snails and slugs feed on living and decayed plants. They make holes in the leaves and can cause stunted growth or even kill a small or weak plant.	• Handpick the pest. • Eliminate decaying plants regularly. Use an organic slug and snail bait with iron phosphate. • Place crushed eggshells and coffee grounds in a ring around plants. • Place orange slices or melon rinds in damp shady places to trap the pests. • Construct a beer trap.
	Spider mites: This pest is very tiny. With a microscope, it can be seen to have four pairs of legs and an oval body. It produces a fine silk webbing on plants.	Spider mites cause tiny white-yellow spots on the leaves and a dusty appearance to start. This progresses to falling pale yellow leaves and a fine webbing over the entire plant.	• Remove the affected leaves and throw them away far from the garden. • Spray with insecticide soap or neem oil. Avoid spraying on a sunny day.
	Squash vine borer: This destructive larva is a cream color with a brown head.	Vine borers feed by tunneling through the stem and destroying the plant tissue. Damage leads to wilting and plant death.	• Although not totally effective, cover the plant until it starts flowering so the borer moth adult can't lay eggs. • Use a Bt (*Bacillus thuringiensis*) injection into the stem. • Keep the garden clean of debris.
	Stink bugs: Stink bugs have a shield-shaped body that is brownish gray.	This pest feeds by piercing plants and sucking the sap from leaves, buds, blossoms, and fruits. This can lead to stunted growth.	• Remove plant debris and keep the garden clean. • Destroy the insect eggs found on the underside of leaves. • Use row covers.
	Tomato and tobacco hornworms: These pests are large caterpillars with eight V-shaped marks on each side. It has a black "horn" on its end.	Hornworms feed on the leaves and fruit of tomato plants. They cause significant damage to the plant, as they are voracious feeders and grow very large. As they feed, they leave behind numerous dark green drops of waste.	• Handpick the caterpillars. Look closely, as their coloration blends in with the leaves. • Spray with Bt (*Bacillus thuringiensis*): Mix 1 tablespoon (15 ml) with 1 gallon (4 L) of water and spray on the top and bottom leaves. • Encourage predatory insects.

VEGETABLE GARDEN DISEASES

DISEASE	SIGN OR SYMPTOM	CONTROL AND PREVENTION
Anthracnose: This disease is caused by a fungus and spreads by spores in water splashed on the leaves.	Anthracnose causes dark lesions on leaves, stems, and fruits, resulting in crop loss.	• Prune infected branches. • Destroy fallen infected leaves. • Avoid overhead watering. • Choose disease-resistant crops and varieties. • Practice crop rotation.
Blossom end rot: This disease is not caused by a fungus; it's an environmental problem. It is most often caused by uneven watering or a calcium deficiency.	Blossom end rot appears as a dark brown sunken area beginning at the blossom end of the fruit.	• Remove the infected fruits. • Fix your plant's calcium levels by fertilizing with bone meal. • Water consistently. • Add a layer of mulch around the plants.
Downy mildew: This disease is caused by a fungus and often occurs during wet weather and moist conditions.	Downy mildew causes yellowish spots on the leaves that progress to a white or gray mold-looking growth on the bottom of leaves.	• Grow disease-resistant cultivars. • Remove the infected leaves or plants. • Prune regularly. • Grow plants in a full sun area. • Avoid getting water on the plants. • Mix 1 tablespoon (14 g) of baking soda with 1 gallon (4 L) of water. Spray the infected area. Avoid spraying in the middle of the day in full sun.
Early blight: This disease of tomatoes and potatoes is caused by a fungus that often spreads during warm, humid conditions.	Early blight happens most often on the oldest leaves at the bottom of the plant. Infected leaves develop small black spots with yellow rings around them. The disease will eventually lead to plant death.	• Remove infected leaves or plants. • Choose disease-resistant cultivars. • Maintain a clean garden by removing plant debris. • Ensure good air circulation around plants. • Avoid overhead watering. • Disinfect your garden tools between plants.
Late blight: Late blight is caused by a water mold called *Phytophthora*, which mostly occurs on tomatoes and potatoes.	Late blight causes large dark brown lesions with a gray-green edge on leaves, fruit, tubers, and stems. The spots enlarge and look oily or water soaked. The disease can rapidly spread to the entire crop.	• Remove infected leaves or plants. • Choose disease-resistant cultivars. • Maintain a clean garden by removing debris. • Ensure good air circulation around plants. • Avoid overhead watering. • Disinfect gardening tools.
Leaf spot: There are many types of leaf spot and they can be caused by fungi, viruses, or bacteria. Leaf spots often appear in humid and wet environments.	Leaf spots are often small yellowish-brown spots on leaves. They can be many different colors, shapes, and sizes that may enlarge, resulting in defoliation of the entire plant.	• Remove infected leaves or plants. • Choose disease-resistant cultivars. • Provide good air circulation around plants. • Avoid overhead watering.

DISEASE	SIGN OR SYMPTOM	CONTROL AND PREVENTION
Potato scab: This common tuber disease is caused by a bacterium-like organism.	Potato scab appears as dark brown patches on the tubers that might also be raised, sunken, rough, and warty.	• Choose disease-resistant cultivars. • Rotate crops and don't plant where potatoes, tomatoes, and other root vegetables have grown within the last 3 years. • Decrease the soil pH level to 5.2 or lower. • Keep the soil moist but do not overwater.
Powdery mildew: This common disease is caused by a fungus that often appears when there isn't enough air circulation between plants. It prefers shade, humidity, and cool nights.	The disease is identified by a white, dusty, powder coating on the leaves, stems, and flowers.	• Choose disease-resistant cultivars. • Remove the infected leaves or plants. • Prune regularly. • Grow plants in a full sun area. • Avoid watering the foliage of the plants. • Mix 1 tablespoon (14 g) of baking soda with 1 gallon (4L) of water. Spray the infected area. Avoid spraying in the middle of a sunny day.
Root rot: Root rot is caused by fungi and water molds in the soil. Growing conditions such as poor soil drainage, overwatering, and excessive soil moisture contribute to the problem.	Root rot signs include wilted plants, yellow leaves, poor growth, thinning plants, and dieback over the entire plant.	• Avoid overwatering plants. • Grow plants in well-drained soil. • Remove highly infected plants to prevent the disease from spreading.
Rust: Rusts are caused by many different species of fungi. Rust spores thrive in moist environments.	Rust starts as white spots on the underside of leaves and on stems. Then, the spots enlarge and become orange and reddish bumps that will turn yellow-green and black.	• Choose disease-resistant cultivars. • Remove the infected leaves or plants. • Ensure good air circulation. • Spray with sulfur- or copper-based fungicides every 7–10 days.
Tomato cracking: Cracks on tomato fruit are caused by too much water or rapid fluctuations in soil water.	Tomato splitting or cracking is usually visible on the upper portions of the fruit.	• Keep a consistent soil moisture. Provide good soil drainage. • Fertilize correctly. • Choose split-resistant varieties. Harvest the unripe tomatoes and ripen them indoors to prevent rotting.
Tomato leaf curl: This problem could be caused by many different reasons: fungus, viruses, improper watering, nutrient deficiencies, insects, and temperature fluctuations.	This problem appears with tomato leaves curling or rolling up. In the beginning, they appear crumpled. Unchecked, tomato leaf curl could lead to stunted growth.	• Choose disease-resistant cultivars. • Ensure tomatoes receive consistent and adequate water.

MINIMIZING PEST AND DISEASE PROBLEMS

↑ Do what you can to minimize problems in your garden by caring for it properly.

- **Grow in Healthy Soil:** Annually, incorporate organic matter into your garden soil to enhance nutrient levels and improve soil structure.
- **Remove and Dispose of Diseased Plants:** Remove the lower leaves of plants that touch the soil. These leaves can easily mold and rot. The lower leaves can also provide easy access for pests to attack the plants. Remove and dispose of diseased plants away from the garden, and do not add them to your compost pile.
- **Rotate Your Crops:** Rotating crops helps to prevent soilborne pest and disease accumulation. Additionally, rotating crops enhances soil fertility by introducing diverse nutrients. Changing crop patterns is advisable to prevent the spread of pests and diseases among crops belonging to the same family. It is recommended to wait at least 2 years before planting the same crop in the same garden location again.
- **Encourage Beneficial Insects:** Introducing beneficial insects such as ladybugs, hoverflies, parasitic wasps, lacewings, solitary bees, butterflies, and ground beetles can be a valuable strategy for pest control in your garden. These insects feed on pests that damage crops and contribute to the overall balance of the ecosystem.
- **Use Companion Planting:** Cultivating a diverse range of crops that mutually benefit one another is a beneficial practice. For instance, grow basil and marigolds alongside tomatoes. Basil has a potent aroma that effectively masks the scent of tomatoes, making it more difficult for pests to locate their favorite host plants.
- **Grow Resistant Varieties:** Resistant plants exhibit a natural ability to withstand the development of specific pests and diseases.
- **Clean Your Garden Tools and Plant Pots:** Regularly cleaning your gardening tools and garden pots is essential to prevent cross-contamination from fungi and other pathogens that can hinder plant growth and development.

FERTILIZING THE GARDEN

Gardening is a multifaceted process that involves the careful cultivation of plants by providing essential elements such as sunlight, fertile soil, and water. The foundation of a healthy and thriving garden lies in your plant roots and soil. Fertilization plays a key role in plant growth by providing essential nutrients. Organic fertilizers such as compost, fish emulsion, manure, bone meal, and worm castings play a crucial role in enriching the soil while supporting soil microbial life, which is crucial for nutrient absorption. It's important to fertilize correctly, as over-fertilizing can lead to nutrient imbalances and harm plant health.

Fish Emulsion

Fish emulsion or fish fertilizer is an organic liquid fertilizer that provides rapid results. Fish fertilizer is particularly rich in nitrogen, but be aware, excessive nitrogen can lead to healthy foliage while hindering fruit and root growth. To apply fish fertilizer, mix 1 to 2 tablespoons (15–30 g) with 1 gallon (4 L) of water and apply it once every 2 to 3 weeks.

Manure

Incorporating manure into the garden can improve soil structure and increase organic matter content. The most commonly used manure sources for vegetable gardens are from chickens, cows, horses, and goats. However, it is crucial to avoid using

↑ There are many product options for fertilizing your vibrant garden.

human, dog, or cat waste in the vegetable garden due to potential health risks.

Worm Castings

Worm castings, a technical term for worm manure, are nutrient-rich fertilizer produced by worms. By consuming organic matter in soil or compost, worms create a valuable resource that enhances soil health, structure, and drainage. Vermicomposting, a home gardening method, facilitates the creation of worm castings. This process not only repurposes food waste and kitchen scraps but also contributes to reducing food waste sent to landfills.

ESTABLISHING A VERMICOMPOSTING SYSTEM

- **Step 1: Prepare the vermicomposting container.** Acquire a wire or plastic bin with numerous holes in the side and bottom and a solid lid. Submerge the container with the lid edge approximately 1 to 2 inches (2.5–5 cm) above the soil level.
- **Step 2: Add composting materials.** Add a layer of coco coir and shredded paper or cardboard to the container. Next, incorporate kitchen and vegetable scraps, which should be chopped into smaller pieces. Finally, add dried leaves.
- **Step 3: Introduce live earthworms.** Introduce live earthworms into the composting mixture. Red wigglers are a good choice.
- **Step 4: Repeat the composting process.** Repeat steps 2 and 3 until the compost reaches the desired height.
- **Step 5: Moisten and seal the container.** Water the compost thoroughly and close the lid. Monitor the compost regularly and adjust the moisture level as needed. Maintain a temperature between 55°F and 75°F (13°C–24°C) for optimal vermicomposting.

WEED MANAGEMENT

Managing weeds in the garden is crucial for growing healthy and productive crops. Weeds are aggressive and invasive plants that quickly grow and compete with vegetables and flowers for nutrients, water, and sunlight—often hindering plant growth. Regular weeding prevents invasive species from spreading and depleting soil resources. Using mulch is a simple yet powerful technique, as mulches not only block sunlight to suppress weeds, but they also retain moisture, and depending on the type, can improve soil health.

↑ Hand pulling weeds is a straight-forward method of removal.

Weed Control Strategies

Hand Pulling

The most effective method for weed removal is hand pulling. This ensures that the entire plant and root system is eliminated. It is recommended to pull weeds when they are small before they produce seeds.

Mulching

Mulching serves as a barrier against weed growth. It also improves soil moisture retention, maintains a clean garden area, and reduces the risk of diseases. Organic mulches include grass clippings, dried leaves, coco coir, and straw. Straw is particularly effective because it decomposes slowly, providing a habitat for beneficial insects and minimizing pest infestations.

Boiling Water

Boiling water is a nontoxic weed control option. Use a tea kettle to boil water and pour it directly on the weeds. Be cautious to avoid burns.

Weed-Resistant Fabric

Fabric that is weed resistant can be used to prevent weed growth in the vegetable planting area. These fabrics block sunlight and allow water to drain through them, making them a porous and effective solution.

WATERING THE GARDEN

Watering is another crucial task in the garden, but it must be done thoughtfully. Vegetables absorb nutrients through their root systems, and nutrients are transported throughout the plant in water solutions. Having a shortage of available water leads to nutrient deficiencies and can cause severe damage to the vegetable plants. Overwatering can lead to root rot, while underwatering can cause wilting and stress. Deep watering encourages healthy root development, and sustainable watering methods like drip irrigation and rainwater harvesting help conserve water. Timing also matters, as watering early in the morning or late into the evening reduces evaporation and ensures plants absorb moisture effectively.

How Much Water Do Plants Need?

Different varieties of vegetables require varying amounts of water, in addition to specific climates and soil characteristics. Consistent rainfall can also reduce water requirements, but it is essential to monitor how much rain does fall in the garden to avoid overwatering, which can be harmful to the plants. A general guideline is that vegetable gardens require approximately 1 inch (2.5 cm) of water per week, but this can be adjusted based on weather conditions and environmental factors. For instance, cucumbers require more water, while herbs thrive with less.

An easy method to determine when your plants require water is to do the finger test. Insert your finger approximately 2 to 3 inches (5–7.5 cm) into the soil. If the soil feels dry, it is an indication that watering is needed.

When Is the Best Time to Water?

Watering in the Morning

The optimal time to water the garden is during the early morning hours when the temperature is still cool. This allows the plants to absorb water effectively and reduces the amount of water that evaporates. Additionally, watering the garden in the morning enables the plant foliage to dry out throughout the day, which helps prevent fungal diseases and other ailments.

Watering in the Middle of the Day

Does watering during midday cause scorching on the foliage? The answer is no. Watering during the hottest part of the day can actually help cool down the plant. However, it is not the most effective method of watering because the water will evaporate quickly before it can reach the roots.

Watering at Night

Watering at night is an optimal time, especially when morning or daytime watering is not feasible. During the night, the weather and soil are cooler, reducing the amount of water evaporation. However, it is crucial to water only at the base of the roots, avoiding the leaves, as water sitting on the leaves can lead to fungal growth like powdery mildew.

↑ Hand watering targets water straight to the roots.

Other Watering Tips

Water at the Base of the Plant
Vegetables obtain water from their roots, so you should concentrate water there and avoid overhead watering on their leaves. Wet leaves, particularly if watering occurs in the evening or during nighttime, can lead to fungal and disease growth. Position the hose at the base of the vegetable and use a lower pressure to minimize the amount of splashing.

Never Overwater
Overwatering also causes problems. Plants require water and oxygen to survive and thrive. Plants dislike soggy feet and excessive water in the soil reduces oxygen availability, leading to plant suffering and eventual death. Signs of overwatering include yellowing leaves, wilting plants, an overly wet soil surface, and an unpleasant soil odor.

Deep Watering
Deeper root systems are associated with healthier plants. Deep watering encourages roots to grow deeper in search of water and nutrients. Additionally, deep watering reduces water loss through evaporation, particularly in hot weather. Generally, in the fall and spring, water your garden once a week, skipping rainy days. In the summer, water your garden twice a week.

↑ Water at the base of the plant to allow the soil to absorb the water for the roots.

PLANT SUPPORT

Plant support refers to structures or tools used to stabilize and guide plants as they grow. These supports are indispensable for maintaining a good, upright plant posture. Supports will safeguard plants from damage due to winds and heavy rainfall, enhance air circulation, and optimize space within the garden. They are particularly beneficial for climbing plants, heavy-fruiting plants, or those with frail stems.

Stakes

Plant stakes are simple, vertical supports used to stabilize plants, keep them upright, and guide their growth. They are particularly useful for young, weak-stemmed, or tall plants prone to bending or breaking under their own weight or due to wind, rain, or heavy fruiting.

There are many different types of plant stakes:

- **Wooden Stakes:** These cost-effective and readily available stakes are constructed from untreated or treated wood. Wooden stakes are an ideal choice for lightweight plants such as tomatoes, peppers, and flowers.
- **Bamboo Stakes:** Lightweight bamboo stakes are crafted from natural bamboo plants, making them an eco-friendly and sturdy staking option. They are available in various lengths and are suitable for supporting vegetables, flowers, and climbing plants such as beans.
- **Metal Stakes:** Constructed from galvanized steel or coated metal, metal stakes are durable, rust resistant, and long lasting. They are suitable for heavier plants such as large tomatoes, sunflowers, or young trees.

- **Plastic or Fiberglass Stakes:** Constructed from synthetic materials, plastic and fiberglass stakes are lightweight, weatherproof, and durable. They are frequently textured for enhanced grip. They are very versatile and appropriate for a diverse range of plants, including small flowers and vegetables.
- **Natural Branches:** DIY stakes can be crafted from trimmed tree branches and are an economical and customizable option for small gardens or rustic garden aesthetics.

Cages

A vegetable cage is a cylindrical or square wire or plastic structure designed to provide all-round support for plants as they grow. They are especially beneficial for bushy, sprawling, or heavy-fruiting plants, as they keep branches off the ground and promote upright growth.

↑ Provide trellising support to plants that need it.

Plant cages are easy to use and can last for multiple growing seasons.

Here are the common types of plant cages:

- **Tomato Cages:** Tomato cages are cone or cylindrical shaped and made from metal or plastic. They are collapsible or stackable for storage, lightweight, and easy to install. They are commonly used to support tomatoes, peppers, and eggplants.
- **Square Cages:** These cages are four-sided structures made of metal or plastic. They offer more space for plant growth compared to the cone-shaped cages. They are suitable for larger plants such as peppers, eggplants, or dwarf fruiting plants.
- **Stackable or Modular Cages:** These vertical or horizontal expandable cages can accommodate various plant sizes with adjustable sections, making them ideal for fast-growing or tall plants such as indeterminate tomatoes and climbing beans.
- **Decorative Plant Cages:** Ornamental cages crafted from wrought iron or painted metal can be functional but harmonize with the aesthetics of the garden. They are commonly used to support flowers such as roses or climbing plants such as scarlet runner beans. These cages not only serve a practical purpose but also contribute to the overall aesthetic appeal of the garden.

CONSERVING WATER IN THE VEGETABLE GARDEN

Here are some practical tips to help in water conservation.

- **Mulch Applications:** Applying an organic mulch effectively helps retain soil moisture.
- **Irrigation Efficiency:** Proper irrigation is the most effective method for water conservation. Set a timer and ensure you provide the appropriate amount of water to the plants. Remember to focus on watering at the base of plants.
- **Rainwater Collection:** Install rain barrels in your garden to collect and store rainwater for use in watering your plants.
- **Graywater:** Using graywater, such as from the kitchen sink and shower (excluding water contaminated with disinfectants, bleach, or grease), is another water conservation strategy that can also reduce water expenses.

4

PRESERVING AND ENJOYING THE HARVEST

The harvest is a time of abundance and rewards, marking the culmination of months of careful gardening and plant care. It is a celebration of nature's cycles, where fruits, vegetables, and herbs reach their peak of ripeness and maturity, offering vibrant colors, rich flavors, and nutritional benefits. For many gardeners, harvesting is about celebrating the effort and patience invested in nurturing plants from seed to maturity. However, the excitement of a successful harvest can often lead to surplus produce that may go to waste if not properly preserved. Effective preservation techniques allow gardeners to extend the enjoyment of their harvest long after the growing season has ended.

Preservation involves various methods to keep produce fresh, flavorful, and safe for extended periods. In the coming section, several popular food preservation techniques will be discussed, including canning, dehydration, pickling, fermenting, and more.

Beyond preservation, enjoying the harvest involves incorporating fresh produce into everyday meals. Garden-fresh vegetables can be used in salads, roasted dishes, stir-fries, and soups, while fruits can be enjoyed raw, blended into smoothies, or baked into desserts. Sharing the harvest with family, friends, and neighbors can strengthen community bonds while encouraging healthier, homegrown eating habits.

Preserving the harvest not only reduces food waste but also promotes sustainability and self-sufficiency. Home-preserved foods often contain fewer preservatives and artificial ingredients compared to store-bought options, supporting healthier diets.

In this section, you'll learn that preserving and enjoying the harvest allows gardeners to celebrate the fruits of their labor while embracing sustainability, healthy eating, and the joy of homegrown produce all year long.

DEHYDRATION

Dehydrating your garden harvest is an effective method for preserving your produce for future consumption. Although food drying has a long history dating back centuries, it has experienced a resurgence in popularity in recent times. Dehydrated foods offer convenience, ease of storage, and an extended shelf life.

Dehydration is a straightforward process that involves the removal of moisture from food to enhance its shelf life. Various methods can be used for dehydration, including the use of a food dehydrator, an oven, or even air-drying certain items.

This comprehensive guide provides instructions on preparing and dehydrating a variety of common fruits, vegetables, and herbs.

Tips for the best dehydration results:

- Harvest produce at its peak ripeness to maximize flavor and nutrient retention.
- Thoroughly wash your harvest to remove dirt and insects. Pat fruits and vegetables dry with a clean towel.
- Slice your fruits and vegetables into even-sized pieces to ensure consistent drying.
- Spread the fruits and vegetables evenly in a single layer on the baking sheet, leaving space between the pieces for good airflow.

Conventional Oven Drying

Oven-based dehydration is a simple and effective method for preserving your harvest without a dedicated dehydrator. The temperature of your oven is crucial for achieving optimal dehydration results. To ensure the best outcome, set your oven temperature below 200°F (93°C).

Typically, dehydrating fruits will take approximately 6 to 12 hours, vegetables 6 to 10 hours, and herbs 1 to 4 hours. It is crucial to check every hour, rotate trays, and flip pieces halfway through to ensure even drying.

Ensure that the dried food cools completely to prevent condensation during storage. Then, store food in airtight containers, glass jars, or vacuum-sealed bags. Place the dehydrated items in a cool, dark, and dry environment for storage.

Electric Dehydrators

Electric dehydrators are among the most efficient and convenient methods used for food drying. These appliances are typically constructed with a timer, temperature gauge, and fans to ensure even heat distribution.

The optimal temperature range for an electric dehydrator to effectively dehydrate fruits and vegetables is from 125°F to 135°F (52°C–57°C). Herbs require a lower temperature range of 95°F to 110°F (35°C–43°C).

The duration depends on the type of food being dehydrated. Generally, fruits require approximately 6 to 24 hours, vegetables 6 to 12 hours, and herbs 1 to 4 hours. It is crucial to check every hour, rotate trays, and flip pieces halfway through to ensure even drying.

Ensure that the dried food cools completely to prevent condensation during storage. Then, store food in airtight containers, glass jars, or vacuum-sealed bags. Store in a cool, dark, and dry environment.

Sun Drying

Sun drying is one of the most ancient and rudimentary methods for food preservation, using only sunlight and warm, dry air. It is particularly suitable for regions with warm climates and low humidity. The most suitable harvests for sun drying include grapes, apricots, figs, plums, cherries, tomatoes, peppers, chilies, onions, basil, thyme, oregano, and mint.

The optimal temperature for sun drying is at least 85°F (29°C) with consistent direct sunlight. However, it is important to note that it may take several days for the plants to fully dry, and the duration can vary depending on the type of food and weather conditions.

FREEZING

Freezing is a straightforward, rapid, and effective method for preserving your garden harvest. Freezing requires minimal supplies and has gained popularity among beginner gardeners. This method effectively preserves the quality, flavor, and nutritional value of food, thereby extending its shelf life. Most food will be maintained at a high quality for 6 to 12 months in the freezer.

This comprehensive guide provides instructions on preparing and freezing a variety of common fruits, vegetables, and herbs.

Tips for the best freezing results:

- Harvest produce at its peak ripeness to maximize flavor and nutrient retention.
- Thoroughly wash your harvest to remove dirt and insects. Pat dry with a clean towel.
- Blanching vegetables before freezing eliminates bacteria, ceases the enzyme activity, and preserves the color, flavor, and texture.
- Effective packaging is crucial to prevent the entry of air into the container or bag, thereby safeguarding moisture retention during freezing.

Blanching

Most veggies should be blanched before freezing. This helps to preserve the flavor, nutrients, and texture of the harvest.

How to blanch vegetables ready to freeze:

- Thoroughly wash the vegetables.
- Boil enough water in a large pot to submerge the vegetables. Gently place the vegetables into the boiling water and cook for 1 to 5 minutes, depending on the vegetable.
- Remove the vegetables from the boiling water and immediately submerge them in an ice bath for an equivalent duration. Drain and pat dry.

Packaging

Effective packaging is crucial in preventing the entry of air into the container or bag. To prevent freezer burn, frozen vegetables should be stored in airtight, rigid containers or heavyweight, moisture-resistant wrap or bags. Most food will be maintained at a high quality for 6 to 12 months in the freezer.

CANNING

Canning is an effective method for preserving food for extended periods, ranging from months to several years. Canning entails sealing food in airtight containers and processing them under controlled conditions to eliminate microorganisms.

There are two widely recognized methods of canning: **water bath canning,** which is suitable for high-acid foods, and **pressure canning,** which is appropriate for low-acid foods.

Water Bath Canning

Water bath canning is an ideal method for preserving high-acid foods such as fruits, tomato sauces, jams, pickles, and salsa. This process is relatively straightforward and does not required too much specialized equipment.

Materials Needed

- Canning jars, also known as Mason jars, with two-piece metal lids (a flat lid and a screw band)
- Water bath canner or a large pot with a rack
- Jar lifter
- Funnel
- Bubble remover, spatula, or chopstick
- Clean kitchen towels

Prepare the Equipment

- Wash the glass jars, lids, and metal bands in hot soapy water.
- Maintain a warm jar temperature to prevent cracking when they are filled with hot food. To do this, you can place the jars in a pot filled with hot water or keep them in a warm oven at 170°F (77°C) until they are filled. Ensure that you follow the manufacturer's instructions for lids. Some require warming; others do not.

Prepare the Food

- Use a reliable canning recipe from a reputable source such as Ball Mason jars or the National Center for Home Food Preservation to guarantee safety and maintain proper acidity levels.
- Use high-acid ingredients or add acidifiers such as bottled lemon juice or vinegar. Never add dairy, flour, rice, pasta, or other thickeners before canning; these can affect safety and must be added later when preparing your meal.
- Select fresh, high-quality produce and other ingredients. Avoid anything overripe, bruised, or damaged.
- Heat the food before packing. Hot packing eliminates air bubbles, reduces spoilage, and ensures proper sealing of the jars.
- If you choose to make raw packs, add raw food to the jars and then pour the hot liquid such as syrup, brine, or juice over it.

Filling Jars

- Ladle hot food into jars using a funnel, ensuring that there is adequate headspace, typically ½ to 1 inch (1–2.5 cm) at the top of the jar.
- Remove any air bubbles from the jars using a bubble remover.
- Wipe the rims of the jars clean with a damp cloth.
- Apply the lids and tighten the screw bands until they are fingertip tight.

Processing the Filled Jars

- Place the sealed jars in boiling water, ensuring that at least 1 to 2 inches (2.5–5 cm) of water surrounds them. Proceed with the processing time as outlined in your recipe, making any necessary adjustments based on your altitude.
- Allow the jars to cool undisturbed for a period of 12 to 24 hours. Subsequently, test the seals by pressing the center of the lid firmly. It should not flex or pop under this pressure. Discard jars with unsealed lids or signs of spoilage, such as cloudy liquid, mold, or unpleasant odors.
- Label the jars with the contents and the date. Store them in a cool, dark, and dry location.

Pressure Canning

Pressure canning is the most reliable method for preserving low-acid foods with a pH > 4.6, including vegetables, meats, poultry, seafood, and certain prepared foods such as soups and stews. In contrast to water bath canning, pressure canning uses elevated temperatures (240°F–250°F [116°C–121°C]) to effectively eliminate hazardous bacteria, such as *Clostridium botulinum* spores, which possess the capability to endure in environments with minimal acidity.

Materials Needed

- Pressure canner: a heavy-duty pot with a locking lid, pressure gauge, and rack
- Canning jars, also known as Mason jars, with two-pieces lids (flat lid and a screw band)
- Jar lifter
- Funnel
- Bubble remover, spatula, or chopstick
- Clean kitchen towels

Prepare the Equipment

- Wash the jars, lids, and bands in hot soapy water.
- Maintain the warm temperature of the jars to prevent cracking when they are filled with hot food. You can place them in a pot with hot water or keep them in a warm oven at 170°F (77°C). Ensure that you follow the manufacturer's instructions for lids. Some require warming; others do not.
- Ensure that the pressure canner lid gasket is intact (if applicable) and that vents and gauges function correctly.

Prepare the Food

- Use a reliable canning recipe from a reputable source such as Ball Mason jars or the National Center for Home Food Preservation to guarantee safety and maintain proper acidity levels.
- Never add dairy, flour, rice, pasta, or other thickeners before canning; these can affect food safety and must be added later when reheating for a meal.
- Select fresh, high-quality produce, meat, or other ingredients. Avoid anything overripe, bruised, or damaged.
- Heat the food before packing. Hot packing eliminates air bubbles, reduces spoilage, and ensures proper sealing of the jars.
- If you choose to make raw packs, add raw food to jars and then pour your hot liquid such as syrup, brine, or juice over it.

Filling Jars

- Ladle the hot food into jars using a funnel, ensuring that there is adequate headspace, typically 1 to 1¼ inches (2.5–4 cm) at the top of the jar.
- Remove any air bubbles from the jars using a bubble remover.
- Wipe the rims of the jars clean with a damp cloth.
- Apply the lids and tighten the screw bands on the jars until they are fingertip tight.

Processing the Filled Jars

- Add 2 to 3 inches (5–7.5 cm) of water to the bottom of the pressure canner.
- Use a jar lifter to place the Mason jars on the rack, ensuring they're not touching.
- Lock the lid in place and leave the vent open.
- Heat the canner until steam escapes steadily from the vent. Let the steam vent for 10 minutes to remove any air pockets.
- Close the vent or add the weighted gauge after venting.
- Heat the canner until it reaches the prescribed pressure. For a dial gauge, the pressure should be 11 PSI at sea level. For a weighted gauge, the pressure should be 10 PSI at sea level (15 PSI for higher altitudes). Begin timing once the pressure is stable.
- Follow the processing time specified in your recipe, which is contingent upon the type of food, the size of the jar, and the altitude at which you are preparing the dish.

- Once the cooking process is complete, turn off the heat and allow the canner to cool naturally. Refrain from opening the lid or venting the steam until the pressure gauge indicates 0. Allow an additional 10 minutes before unlocking the lid.
- Allow the jars to cool undisturbed for a period of 12 to 24 hours. Subsequently, test the seals by pressing the center of the lid firmly. It should not flex or pop under this pressure. If the jar did not seal, refrigerate it and use the contents within 1 week.
- Label the jars with the contents and the date. Store them in a cool, dark, and dry location for up to 1 to 2 years.

FERMENTATION

Fermentation is a natural metabolic process where microorganisms such as bacteria, yeast, and molds break down organic compounds, primarily sugars, into simpler substances like acids, gases, or alcohol. This ancient preservation technique has been used for thousands of years to enhance food shelf life, improve flavor, and increase nutritional value.

Fermentation can be used for both food and beverages, contributing to the creation of products like yogurt, sauerkraut, kimchi, cheese, sourdough bread, kombucha, beer, and wine. Beyond preservation, fermentation also plays a crucial role in the development of complex flavors and the production of probiotics that contribute to gut health. Furthermore, fermentation can increase the bioavailability of nutrients, meaning the body can absorb vitamins and minerals more efficiently. Here's a detailed step-by-step guide to the fermentation process.

Select High-Quality Ingredients

The foundation of successful fermentation starts with fresh, high-quality ingredients. Choose organic, pesticide-free produce, as synthetic chemicals can interfere with the natural microbial activity required for fermentation. Avoid heavily washed or treated produce, as natural bacteria on the surface aids in fermentation.

Fermentation ingredients to use:

- **Vegetables:** Cabbages, cucumbers, carrots, radishes, beets
- **Fruits:** Apples, grapes, pears, berries
- **Dairy:** Raw or pasteurized milk for yogurt, kefir, and cheese
- **Grains and Legumes:** Barley for beer and soybeans for miso and tempeh
- **Sweet Liquids:** Fruit juices, honey, or sugar water for alcoholic fermentation like wine or mead

Prepare the Ingredients

Proper preparation ensures that food ferments evenly and safely.

- Wash and rinse food thoroughly with clean water to remove any dirt and debris while preserving the natural microbes.
- Cut and slice the food. Shred cabbage for sauerkraut and slice cucumbers for pickles.
- Bruise or crush the food. Bruising can help release natural sugars, promoting enhanced microbial activity—like crushing grapes for wine production.
- Add salt. For lacto-fermentation, use noniodized salt, as iodine can inhibit beneficial bacteria.

Prepare the Fermentation Medium

Depending on the type of fermentation, you may need to prepare a brine, starter culture, or sugar solution.

These are different types of fermentation mediums:

- **Brine for Vegetables:** Dissolve noniodized salt in water. Use a 2% to 3% salt solution or put about 1 to 2 tablespoons (18–36 g) of salt per 1 quart (1 L) of water.
- **Starter Cultures:** For yogurt and cheese, you will need to add a bacterial starter culture or previously fermented yogurt.
- **Sugary Liquids:** For mead or fruit wines that require an alcoholic fermentation, dissolve sugar or honey in water to make a fermentation base.

Pack Ingredients into the Fermentation Vessel

Packing food correctly is crucial for a successful fermentation.

- **Use Clean Containers:** Use sterilized glass jars, ceramic crocks, or food-grade plastic containers for fermenting.
- **Pack Tightly:** After filling, press the food ingredients down firmly to remove any air pockets in the jars to ensure even fermentation. Make sure the food is fully submerged in the brine (or its own juices) to prevent mold growth.

Create the Ideal Fermentation Environment

Microorganisms thrive under specific conditions. Different types of fermentation can be done with or without air.

- **Anaerobic (Without Oxygen) Fermentation:** For lacto-fermentation, as with sauerkraut, the food must stay fully submerged and sealed under the liquid to prevent oxygen exposure, which can encourage mold. Use fermentation weights or plates to hold down the food items.
- **Aerobic (with Oxygen) Fermentation:** For kombucha and sourdough starters, cover the container with a breathable cloth to allow oxygen to circulate while preventing contaminants from entering. Store the fermentation containers in a dark or dimly lit space to prevent degradation and excessive heat.

Start the Fermentation Process

Once packed and sealed, fermentation will begin as beneficial microorganisms break down the sugars into acids, alcohol, and gases.

- **Lacto-Fermentation:** In lacto-fermentation, beneficial bacteria convert sugars into lactic acid, lowering the pH and preserving the food.
- **Alcohol Fermentation:** Alcohol fermentation involves yeast converting sugar into alcohol and carbon dioxide. This type of fermentation is used in wine, beer, and bread making.
- **Acetic Acid Fermentation:** In acetic acid fermentation, alcohol is converted into vinegar through exposure to oxygen and acetic acid bacteria.

Monitor the Ferment Regularly

Changes during the fermentation process can be good or bad. Be sure to monitor frequently. These are some changes that might occur:

- **Bubbling:** An active fermentation produces carbon dioxide, which is visible as bubbles.
- **Odor:** A sour, tangy, or yeasty smell is normal. But watch for foul, rotten, or ammonia-like odors, which can indicate spoilage.
- **Color:** The brine may turn cloudy, which is normal. However, a blue, pink, or fuzzy mold is a sign of spoilage.
- **Taste:** Carefully taste the ferment as it progresses. The flavors should deepen with time.

Determine When the Fermentation Is Complete

The fermentation time varies depending on the type of food and environmental conditions. The signs of completion include a sour taste, reduced bubbling, and clear brine or liquid.

Typical fermentation times for various items:

- **Sauerkraut and Kimchi:** 1 to 4 weeks
- **Pickles:** 6 to 12 hours
- **Yogurt:** 6 to 12 hours
- **Sourdough:** 5 to 7 days
- **Wine and Beer:** 2 to 4 weeks (then aging)

Storage After Fermentation

- **Cold Storage:** Transfer the finished fermented product to a refrigerator or root cellar (34°F–50°F [1°C–10°C]) to slow microbial activity and preserve the food for months. Most fermented foods are best stored this way.
- **Room Temperature Storage:** Some fermented pickled vegetables, kept in brine, can be stored in a cool, dark location. For maximum crunch, eat them quickly. And some fermented liquids, like beer and wine, are best stored in cool, dark pantries to age.

Fresh Herb Seasoned Salt

I love making homemade seasoned salts that are infused with fresh herbs from my garden. Their captivating aroma is remarkable. Fresh herb seasoned salt is a simple and delicious way to preserve the essence of culinary treasures from your garden—it is versatile and can be used for cooking, grilling, and serving.

INGREDIENTS

- 1 cup (150 g) kosher salt, sea salt, or coarse salt
- ½ cup (30 g) fresh herbs such as rosemary, thyme, basil, parsley, dill, or oregano
- 1 to 2 cloves garlic, optional
- Lemon zest, optional

STEP 1. Prepare the herbs by thoroughly washing them with clean water. Pat them dry with a clean towel. Remove the tough stems, retaining only the tender leaves.

STEP 2. Combine the salt and herbs in a food processor. Pulse until the herbs are finely chopped and evenly distributed throughout the salt. Alternatively, if you lack a food processor, you can manually chop the herbs finely with a knife and mix them into the salt.

STEP 3. Dry the mixture using your method of choice.

Oven Method: Spread the herb and salt mixture thinly on a baking sheet. Bake at 170°F (77°C) or at the lowest oven setting for 1 to 2 hours, stirring occasionally, until it is completely dry.

Dehydrator Method: Spread the mixture on a dehydrator tray. Dry at 95°F to 115°F (35°C–46°C) for 1 to 2 hours or until dry.

STEP 4. Once the mixture is dry, break up any clumps and transfer the herb salt to a clean, airtight container or glass jar. Label with the date and herbs used.

STEP 5. Store in a cool, dark, and dry place. Herb seasoned salt can last up to 6 months.

The Best Chili Oil

Chili oil enhances the flavor and overall taste of many dishes. It has become a staple condiment in my household, and my children and family members enjoy its versatility. Making it at home allows you to customize the spice level and flavors to suit your taste. This chili oil recipe has gained significant popularity on my Instagram page.

INGREDIENTS

- 4 to 6 cups (946–1420 ml) vegetable oil, plus more oil as needed
- 1 cup (150 g) finely chopped garlic
- 1 cup (160 g) finely chopped shallots
- 2 cups (115 g) Thai ground chili flakes
- 1 cup (58 g) Korean (gochugaru) red chili flakes
- 1 cup (275 g) chili garlic sauce
- 1 cup (235 ml) soy sauce
- ½ cup (100 g) sugar
- ½ cup (120 ml) chicken marinade sauce, optional
- ½ cup (72 g) toasted sesame seeds

Mixed Pickled Vegetables

My all-time favorite pickled vegetables are the ones my mother used to make when I was growing up. These tangy, slightly sweet, and crunchy condiments are made from a combination of fresh vegetables preserved in a vinegar-based brine. This colorful side dish is a staple in Cambodian cuisine, often served alongside grilled meats, rice dishes, and noodle bowls to balance the rich and savory flavors. In contrast to fermented pickles, this quick vegetable pickle can be consumed within hours of preparation.

INGREDIENTS

- 1 cup (130 g) julienned or chopped carrots
- 1 cup (110 g) julienned or chopped daikon radish
- 1 cup (119 g) cucumber, sliced ¼ inch (0.6 cm) thick
- 1 cup (100 g) small florets cauliflower
- 1 cup (90 g) chopped cabbage
- 3 tablespoons (54 g) salt, divided
- 1 cup (235 ml) white vinegar or rice vinegar
- 1 cup (235 ml) water
- ½ cup (100 g) sugar

STEP 1. Prepare the vegetables. Wash and cut up all the vegetables into bite-sized pieces. Sprinkle 1 tablespoon (18 g) of salt over the vegetables and toss well. Let them sit for 20 to 30 minutes to draw out their moisture and slightly soften. Rinse the vegetables well with cold water and drain thoroughly.

STEP 2. Make the pickle brine. In a saucepan, combine vinegar of choice, water, sugar, and the remaining 2 tablespoons (36 g) salt. Heat the mixture over medium heat, stirring until the sugar and salt are dissolved completely. Remove the brine from the heat and let it cool to room temperature.

STEP 3. Pack the vegetables. Place the mixed vegetables into clean, sterilized glass jars. Pour the cooled pickling brine over the vegetables, ensuring they are fully submerged, and seal the jars tightly. Let the jars sit at room temperature for a couple of hours before serving for mild pickles. For a tangier flavor, refrigerate for 2 to 3 more days before eating.

STEP 4. For optimal freshness, it is recommended to consume the vegetable pickles within a 1-to-2-week period. However, they can be stored for an extended duration if maintained in a properly sealed and refrigerated container.

STEP 1. Pour the oil into a large wok and heat it until it is hot. You can drop a piece of garlic into the oil to test its temperature.

STEP 2. Add the chopped garlic and shallots and cook until they have turned a slightly golden color.

STEP 3. Reduce the heat to medium-low, then add the Thai ground chili flakes, gochugaru, and chili garlic sauce. Stir until the ingredients are well combined. Cook for about 1 to 2 minutes.

STEP 4. Add the soy sauce, sugar, and chicken marinade sauce, if using. Cook for a couple more minutes.

STEP 5. Turn off the heat and add the toasted sesame seeds.

Fresh Kimchi Without Fermentation

Fresh kimchi (called *geotjeori* in Korean) is a nonfermented variation of traditional kimchi. It is prepared in a similar manner but can be consumed immediately after mixing, providing a refreshing, crunchy, and mildly spicy flavor. In contrast to regular kimchi, which undergoes a fermentation process to develop tangy and sour flavors, fresh kimchi retains the natural sweetness and crispness of the vegetables.

I personally prefer making fresh kimchi because it preserves the freshness of homegrown vegetables from the garden, and it is quite straightforward to prepare. It is commonly served as a side dish with Korean BBQ, rice, noodles, or stews. It is an ideal choice for individuals seeking the taste of kimchi without the necessity of waiting for fermentation.

Fresh kimchi is most flavorful and crisp when consumed within 5 to 7 days of its production. As time passes, it may undergo slight fermentation, resulting in a tangier flavor profile.

INGREDIENTS

- 1 medium napa cabbage
- ¼ cup (72 g) sea salt
- ½ cup (29 g) gochugaru (Korean red pepper flakes)
- 5–7 garlic cloves, minced
- 2 teaspoons (16 g) grated ginger
- 1 tablespoon (15 ml) fish sauce*
- 1½ tablespoons (25 ml) soy sauce*
- 1–2 tablespoons (13–26 g) sugar
- 1–2 tablespoons (15–28 ml) sesame oil
- 4 to 5 stems green onions or garlic scapes, cut into 1-inch (2.5 cm) pieces
- ¾ cup (98 g) julienned carrot
- ¾ cup (83 g) julienned daikon radish
- 1 tablespoon (8 g) toasted sesame seeds

* For a vegetarian option, use 2½ tablespoons (40 ml) soy sauce and omit the fish sauce.

STEP 1. Prepare the napa cabbage by cutting it into bite-sized pieces. Rinse thoroughly in cold water and drain. Sprinkle the sea salt evenly over the cabbage and toss well. Note: I skip the salting step for homegrown napa cabbage. Let it sit for 20 to 30 minutes to draw out the moisture. Toss the cabbage every 15 minutes for even salting. Rinse the cabbage thoroughly 2 to 3 times in cold water to remove the excess salt. Drain well and set aside.

STEP 2. Make the kimchi sauce. In a bowl, combine the gochugaru, garlic, ginger, fish sauce, soy sauce, sugar, and sesame oil. Mix well into a thick paste.

STEP 3. Mix the kimchi together. In a large mixing bowl, combine the napa cabbage, green onions or garlic scapes, carrots, radish, and toasted sesame seeds. Pour the kimchi sauce over the vegetables. Using gloves to avoid staining your hands, gently massage the sauce into the vegetables until everything is evenly coated.

STEP 4. Serve immediately or leave it in the refrigerator for a couple hours before serving.

STEP 5. Store leftover kimchi in an airtight container in the refrigerator for up to 1 week to preserve its optimal flavor and crispness.

Homemade Canned Tomato Sauce

Homemade tomato sauce from the garden is a wholesome, vibrant, and wildly customizable kitchen staple. It is a rich, flavorful sauce made using freshly harvested tomatoes, offering a taste of peak-season produce. This sauce captures the natural sweetness and acidity of vine-ripened tomatoes, making it ideal for pasta, pizza, soups, and more.

INGREDIENTS

- 10 pounds (5 kg) fresh tomatoes
- 2 teaspoons (36 g) salt
- 1 teaspoon (4 g) sugar
- 1 teaspoon (4.5 g) dried basil, oregano, and thyme, optional
- ¼ cup (60 ml) bottled lemon juice

STEP 1. Prepare the tomatoes. Wash tomatoes thoroughly. Boil a large pot of water and prepare an ice water bath. Score an "X" on the bottom of each tomato and drop them into the boiling water for about 40 to 60 seconds. Transfer them immediately into the ice bath and the skins should peel off easily. Core and chop the peeled tomatoes.

STEP 2. Cook and blend the sauce. Put the chopped tomatoes in a large pot over medium heat. Simmer for 30 minutes until softened. Blend the tomatoes using a food mill for a smooth sauce or an immersion blender for a chunkier texture. Add salt, sugar, and herbs if using. Simmer the sauce for another 30 to 60 minutes, allowing it to reduce and thicken.

STEP 3. Prepare the jars and the water bath. Sterilize canning jars and lids by boiling them in water for 10 minutes. Fill a large water bath canner with enough water to cover the jars by 1 to 2 inches (2.5–5 cm) and bring it to a simmer.

STEP 4. Fill the jars. Add 1 tablespoon (15 ml) bottled lemon juice per pint jar or 2 tablespoons (30 ml) per quart jar to ensure a safe acidity level. DO NOT skip adding the lemon juice. Ladle the hot tomato sauce into jars, leaving ½ inch (1 cm) headspace. Using a bubble remover, remove the air bubbles and wipe the jar rims clean. Place the sterilized lids on the jars and twist on the bands until finger tight.

STEP 5. Process the jars. Place the jars in the boiling water bath canner. Ensure they are fully submerged with 1 to 2 inches (2.5–5 cm) of water covering the tops. Boil for the following times, adjusting for altitude if necessary.

Pint Jars: 35 minutes
Quart Jars: 40 minutes

STEP 6. Cool and store the jars. Remove the jars carefully using a jar lifter and place them on a towel to cool undisturbed for 12 to 24 hours. Check the seals by pressing the center of each lid. If it doesn't flex, the seal is secure. Store the jars in a cool, dark place for up to 12 months.

ABOUT THE AUTHOR

Sandra Mao is the creator and founder of the brand Sandra Urban Garden. She lives in Southern California, USDA Hardiness Zone 10B. Sandra has a BA in business management and interior design, but she has always been passionate about vegetable gardening. She started to grow fruit trees and other edibles when she moved to California. Because of the blessings of the weather and lack of freezing temperatures, she is able to grow most vegetables and fruits year-round, though her content is aimed at gardeners from all growing zones. Sandra first started growing her edible garden in containers and has expanded the growing space over the years. She loves helping new gardeners start growing their own food by sharing her passion for growing colorful and productive varieties, offering unique tips and tricks, and teaching others how to use the harvest wisely. Follow along on Sandra's gardening journey: @sandra.urbangarden

ACKNOWLEDGMENTS

This book would not have been possible without the unwavering support, love, and encouragement of my family and the incredible efforts of the teams who helped bring it to life.

To my grandmother, mother, and aunt, thank you for your unconditional love. Thank you for inspiring my love of gardening. From the time I was young, you taught me the joy of nurturing plants and the patience required to see them grow. Your wisdom and gentle touch in the garden left an incredible mark on me. I recall learning to protect fruits from birds by wrapping them in paper. Every morning, I was excited to pick jasmine for my grandmother. I treasure the way you made every flower, vegetable, and herb feel special, and how you taught me that gardening is a process of learning, experimenting, and nurturing with love.

To my beloved husband, Mora, thank you for your endless love, unwavering support, and boundless patience throughout this journey. Your belief in me, even when I doubted myself, has been a constant source of strength. You have been my rock, my sounding board, and my biggest cheerleader. I am forever thankful!

To my children, Darveedt, Kenneth, and Lauren, thank you for your love, patience, and understanding as I embarked on this journey. Your curiosity, joy, and enthusiasm have inspired me every day, whether you were lending a helping hand in the garden, offering encouragement, or simply reminding me of the beauty in the little things.

To my nephew, Preston, thank you for being such an incredible model. Your willingness to step in, your natural talent, and your ability to bring energy and life to each shot made all the difference.

To Jessica, Winnie, Regina, John, and the Quarto teams, thank you for your expertise, dedication, and hard work throughout the process of bringing this book to life. From the initial concept to the final product, your support has been indispensable.

To my dear friend Resh, thank you for connecting me with the publisher who made this book possible. Without your support and guidance, I may never have had the opportunity to work with such a talented and professional team. I am grateful for your friendship.

To Kami, thank you for your incredible work and your skill behind the lens, your professionalism, and the unique perspective you brought to this project.

To Vego Garden, thank you for your support and commitment to quality and sustainability. Your products have been a cornerstone in bringing Sandra Urban Garden to life. The raised beds and gardening products played a crucial role in creating a thriving vibrant garden space and I'm grateful for your support.

To my social media friends and followers, thank you for your unwavering support and encouragement throughout this journey. Your kind words, thoughtful comments, and continued enthusiasm have been a constant source of motivation. Whether you've shared your own experiences, offered advice, or simply followed along, your presence has made this project even more meaningful. I am deeply grateful to each one of you, and your belief in this work has made all the difference. This book exists because of you! I am honored to have such a wonderful community behind me.

INDEX